Abiotic and Biotic Stress in Plants

Edited by Alexandre Bosco de Oliveira

Published in London, United Kingdom

IntechOpen

Supporting open minds since 2005

Abiotic and Biotic Stress in Plants
http://dx.doi.org/10.5772/intechopen.77845
Edited by Alexandre Bosco de Oliveira

Contributors
Mukesh K. Kumar Berwal, Chet Ram, Puran Bridgemohan, Majeed Mohammed, Ertan Yıldırım, Raziye Kul, Aslıhan Esringü, Esin Dadaşoğlu, Üstün Şahin, Metin Turan, Selda Örs, Güleray Agar, Melek Ekinci, Szilvia Veres, László Zsombik, Csaba Juhász, Faisal Hussain, Farzana Usman, Ajaz A. Lone, Audil Gull, Peter Kaloki, Viola Devasirvatham, Daniel K. Y. K. Y. Tan, Gideon Onyekachi Okoro, Ogbonnaya Boniface Onu, Felix Ngasoh, Nicholas Namessan, Bimal Das, Aparajita Das, Karaba N. Nataraja, Raju Y. Soolanayakanahally, Karen K. Tanino, Tawhidur Rahman, Kunhikrishnan H. Dhanyalakshmi

First published in London, United Kingdom, 2019 by IntechOpen
IntechOpen is the global imprint of INTECHOPEN LIMITED, registered in England and Wales, registration number: 11086078, The Shard, 25th floor, 32 London Bridge Street
London, SE19SG – United Kingdom
Printed in Croatia

British Library Cataloguing-in-Publication Data
A catalogue record for this book is available from the British Library

Additional hard and PDF copies can be obtained from orders@intechopen.com

Abiotic and Biotic Stress in Plants
Edited by Alexandre Bosco de Oliveira
p. cm.
Print ISBN 978-1-78923-811-2
Online ISBN 978-1-78923-812-9
eBook (PDF) ISBN 978-1-83962-485-8

For EU product safety concerns:
IN TECH d.o.o., Prolaz Marije Krucifikse Kozulić 3, 51000 Rijeka, Croatia, info@intechopen.com or visit our website at intechopen.com.

Meet the editor

Alexandre Bosco de Oliveira has a BS in Agronomic Engineering (2006) and Secondary Education in Biological Sciences (2009). He obtained his MS (2008) and PhD (2010) in Plant Science at the Federal University of Ceará-UFC, Brazil, and worked as a visiting researcher at the University of Florida (2017), USA. His academic background involves research activities with biotic and abiotic stresses in plants, agriculture, and crop physiology. In 2012, he started his current position as Professor of Agriculture at UFC. Since then, he has become the Head of the Crop Physiology Laboratory and coordinates the Ecophysiology of Semiarid Agroecosystems Research Group, advising undergraduate and graduate students. To date, he has published six books, 21 chapters, 66 articles in peer-reviewed journals, and contributes as a reviewer and associate editor to scientific journals worldwide.

Contents

Preface

Plants are subjected to numerous environmental stresses, which can be classified into two broad areas: abiotic and biotic stresses. While the first is considered the damage done to an organism by other living organisms, the latter occurs as a result of a negative impact of non-living factors on the organisms. In this scenario, the current most accepted opinion of scientists is that both biotic and abiotic factors in nature and agroecosystems are affected by climate change, which may lead to significant crop yield decreases worldwide. We should take into consideration not only this environmental concern but also the fact that 20 years from now the earth's population will need 55% more food than it can produce now. Therefore, it is crucial to address such concerns and bring about possible solutions to future plant stress-related outcomes that might affect global agriculture.

Facing the increasing population, and consequently higher demand for food, fiber, and biofuels, the demand for agricultural products is set to increase in the next few years. Thus, plant science research to improve sustainable production worldwide is the vital task of the scientific community to address the increasing demands and needs for such products. In the last few decades, the results of wide-scale research have been applied, field mechanization improved, technical and agricultural methods developed, research and training extended, and new methods established worldwide.

This book intends to provide the reader with a comprehensive overview of both biotic and abiotic stresses through 10 chapters that include case studies and literature reviews about these topics. Hence, this volume presents outstanding chapters involving theoretical and practical research work carried out by experienced researchers. There will be a particular focus on understanding the physiological, biochemical, and molecular changes observed in stressed plants as well as the mechanisms underlying stress tolerance in plants.

Taking into consideration the fact that environmental vulnerabilities are the significant constraints for growth, development, and productivity of crops, this book is a must-read work for academic students, scientists, and enthusiasts about this theme. The methods and technologies recommended here have resulted in essential highlights about how to tackle abiotic/biotic stress in plants, achieve higher yields, and maximize the use of inputs under harmful environmental conditions. In other words, this is a valuable resource about biotic and biotic factors that affect crops' yields, particularly for those who work in research organizations and higher academic institutions. This textbook is a precise and complete work that will benefit graduates, postgraduates, and researchers who work with environmental stressors in plants.

As editor, I am grateful to all the authors who have written their chapters meticulously and contributed their valuable work to this book. I would also like to thank the editorial staff of IntechOpen Publisher and its team for all the kind support provided throughout the whole editorship process, enabling this book to be produced on time and in an excellent manner. I express my special thanks to my mother

Francisca, my wife Maria, and my kids Matheus and Giovana, for inspiring me and being my pillars of strength. Last but not least, my deepest gratitude is for my Lord and Savior, Jesus Christ, who takes care of me and gives me health to make my dreams come true. "I will give thanks to you, LORD, with all my heart; I will tell of all your wonderful deeds." Ps. 9:1.

Alexandre Bosco de Oliveira
Professor of Agriculture,
Plant Science Department,
Federal University of Ceará,
Fortaleza, Ceará, Brazil

Section 1

Multiple Stresses

Chapter 1

Biotic and Abiotic Stresses in Plants

Audil Gull, Ajaz Ahmad Lone and Noor Ul Islam Wani

Abstract

Plants are subjected to a wide range of environmental stresses which reduces and limits the productivity of agricultural crops. Two types of environmental stresses are encountered to plants which can be categorized as (1) Abiotic stress and (2) Biotic stress. The abiotic stress causes the loss of major crop plants worldwide and includes radiation, salinity, floods, drought, extremes in temperature, heavy metals, etc. On the other hand, attacks by various pathogens such as fungi, bacteria, oomycetes, nematodes and herbivores are included in biotic stresses. As plants are sessile in nature, they have no choice to escape from these environmental cues. Plants have developed various mechanisms in order to overcome these threats of biotic and abiotic stresses. They sense the external stress environment, get stimulated and then generate appropriate cellular responses. They do this by stimuli received from the sensors located on the cell surface or cytoplasm and transferred to the transcriptional machinery situated in the nucleus, with the help of various signal transduction pathways. This leads to differential transcriptional changes making the plant tolerant against the stress. The signaling pathways act as a connecting link and play an important role between sensing the stress environment and generating an appropriate biochemical and physiological response.

Keywords: environmental stresses, temperature, radiation, bacteria, fungi, cellular response and signaling

1. Introduction

Stress in plants refers to external conditions that adversely affect growth, development or productivity of plants [1]. Stresses trigger a wide range of plant responses like altered gene expression, cellular metabolism, changes in growth rates, crop yields, etc. A plant stress usually reflects some sudden changes in environmental condition. However in stress tolerant plant species, exposure to a particular stress leads to acclimation to that specific stress in a time time-dependent manner [1]. Plant stress can be divided into two primary categories namely abiotic stress and biotic stress. Abiotic stress imposed on plants by environment may be either physical or chemical, while as biotic stress exposed to the crop plants is a biological unit like diseases, insects, etc. [1]. Some stresses to the plants injured them as such that plants exhibit several metabolic dysfunctions [1]. The plants can be recovered from injuries if the stress is mild or of short term as the effect is temporary while as severe stresses leads to death of crop plants by preventing flowering, seed formation and induce senescence [1]. Such plants will be considered to be stress susceptible. However several plants like desert plants (Ephemerals) can escape the stress altogether [2].

Biotic stress in plants is caused by living organisms, specially viruses, bacteria, fungi, nematodes, insects, arachnids and weeds. The agents causing biotic stress directly deprive their host of its nutrients can lead to death of plants. Biotic stress can become major because of pre- and postharvest losses. Despite lacking the adaptive immune system plants can counteract biotic stresses by evolving themselves to certain sophisticated strategies. The defense mechanisms which act against these stresses are controlled genetically by plant's genetic code stored in them. The resistant genes against these biotic stresses present in plant genome are encoded in hundreds. The biotic stress is totally different from abiotic stress, which is imposed on plants by non-living factors such as salinity, sunlight, temperature, cold, floods and drought having negative impact on crop plants. It is the climate in which the crop lives that decides what type of biotic stress may be imposed on crop plants and also the ability of the crop species to resist that particular type of stress. Many biotic stresses affect photosynthesis, as chewing insects reduce leaf area and virus infections reduce the rate of photosynthesis per leaf area.

Abiotic stresses such as drought (water stress), excessive watering (water logging), extreme temperatures (cold, frost and heat), salinity and mineral toxicity negatively impact growth, development, yield and seed quality of crop and other plants. In future it is predicted that fresh water scarcity will increase and ultimately intensity of abiotic stresses will increase. Hence there is an urgency to develop crop varieties that are resilient to abiotic stresses to ensure food security and safety in coming years. A plants first line of defense against abiotic stress is in its roots. The chances of surviving stressful conditions will be high if the soil holding the plant is healthy and biologically diverse. One of the primary responses to abiotic stress such as high salinity is the disruption of the Na^{+}/K^{+} ratio in the cytoplasm of the plant cell. The phytohormone abscisic acid (ABA) plays an important role during plant adaptation to environmental stress such as high salinity, drought, low temperature or mechanical wounding [3].

2. Crop plants and abiotic stresses

Plants are encountered by number of abiotic stresses which impact on the crop productivity worldwide. These abiotic stresses are interconnected with each other and may occur in form of osmotic stress, malfunction of ion distribution and plant cell homeostasis. The growth rate and productivity is affected by a response caused by group of genes by changing their expression patterns. So, the identification of responsive genes against abiotic stresses is necessary in order to understand the abiotic stress response mechanisms in crop plants. The abiotic stresses occurring in plants include.

2.1 Cold

Cold stress as abiotic stress has proved to be the main abiotic stresses that decrease productivity of agricultural crops by affecting the quality of crops and their post-harvest life. Plants being immobile in nature are always busy to modify their mechanisms in order to prevent themselves from such stresses. In temperate conditions plants are encountered by chilling and freezing conditions that are very harmful to plants as stress. In order to adopt themselves, plants acquire chilling and freezing tolerance against such lethal cold stresses by a process called as acclimation. However many important crops are still incompetent to the process of cold acclimation. The abiotic stress caused by cold affect the cellular functions of plants in every aspect. Several signal transduction pathways are there by which these cold stresses are transduced like components of ROS, protein kinase, protein phosphate, ABA and Ca^{2+}, etc. and among these ABA proves to be best.

2.2 Salt

Soil salinity poses a global threat to world agriculture by reducing the yield of crops and ultimately the crop productivity in the salt affected areas. Salt stress reduces growth of crops and yield in many ways. Two primary effects are imposed on crop plants by salt stress; osmotic stress and ion toxicity. The osmotic pressure under salinity stress in the soil solution exceeds the osmotic pressure in plant cells due to the presence of more salt, and thus, limits the ability of plants to take up water and minerals like K^+ and Ca^{2+}. These primary effects of salinity stress causes some secondary effects like assimilate production, reduced cell expansion and membrane function as well as decreased cytosolic metabolism.

2.3 Drought

Nowadays climate has changed all around the globe by continuously increase in temperature and atmospheric CO_2 levels. The distribution of rainfall is uneven due to the change in climate which acts as an important stress as drought. The soil water available to plants is steadily increased due severe drought conditions and cause death of plants prematurely. After drought is imposed on crop plants growth arrest is the first response subjected on the plants. Plants reduce their growth of shoots under drought conditions and reduce their metabolic demands. After that protective compounds are synthesized by plants under drought by mobilizing metabolites required for their osmotic adjustment.

2.4 Heat

Increase in temperature throughout the globe has become a great concern, which not only affect the growth of plants but their productivity as well especially in agricultural crops plants. When plants encounter heat stress the percentage of seed germination, photosynthetic efficiency and yield declines. Under heat stress, during the reproductive growth period, the function of tapetal cells is lost, and the anther is dysplastic.

2.5 Toxin

The increased dependence of agriculture on chemical fertilizers and sewage waste water irrigation and rapid industrialization has added toxic metals to agriculture soils causing harmful effects on soil-plant environment system.

3. Crop plants and biotic stresses

Plants struggle with many kinds of biotic stresses caused by different living organisms like fungi, virus, bacteria, nematodes, insects etc. These biotic stress agents cause various types of diseases, infections and damage to crop plants and ultimately affect the crop productivity. However, different mechanisms have been developed through research approaches to overcome biotic stresses. The biotic stresses in plants can be overcome by studying the genetic mechanism of the agents causing these stresses. Genetically modified plants have proven to be the great effort against biotic stresses in plants by developing resistant varieties of crop plants.

4. Polyamine: plant response to stresses

Plants being immobile in nature have to go through continuous fluctuations in the environment with appropriate physiological, developmental and biochemical changes [4]. More than 50% reduction in crop plants occur due to abiotic stresses worldwide which is the main cause of crop loss [5]. To counteract the stresses, plants are equipped with a large set of defense mechanisms [6]. Among the different classes of compatible solutes, polyamines stand as one of the most effective against extreme environmental stress. Polyamines are low molecular weight aliphatic nitrogen compounds positively charged at physiological pH [7]. Investigations into plant polyamines at a molecular level have led to isolation of a number of genes encoding polyamine biosynthetic enzymes from a variety of plant species [8]. In recent years, molecular and genomic studies with mutants and transgenic plants having no or altered activity of enzymes involved in the biosynthesis of polyamines have contributed to a better understanding of biological functions of polyamines in plants.

4.1 Polyamine and plant response to abiotic stresses

Stress derived changes in cellular polyamines provide clues on their possible implication in stress but do not provide evidence of their role in counteracting stress. The levels of endogenous polyamines can be increased by application of exogenous polyamines, which has been attempted before or during stress [9, 10]. Exogenous application of polyamines could preserve plant cell membrane integrity, minimize growth inhibition caused by stress, moderate expression of osmotically responsive genes and increase activities of antioxidant enzymes. In another approach treatment with biosynthesis inhibitors can reduce endogenous polyamine resulted in stress sensitive phenotypes. However this effect is reversed by the concomitant application of exogenous polyamine [9, 11]. Another genetic approach employed for analyzing biological functions of polyamine metabolism in stress response is the use mutant deficient in polyamine biosynthesis [12].

4.2 Polyamine and plant response to biotic stresses

Polyamine metabolism has long been known to distort in plant cells responding to insightful changes in plants interacting with fungal [13], viral pathogens [14] and mycorrhizae [15]. It is hard to identify the contribution of polyamine accumulation in infected organs as it is present both in plants and pathogenic fungi. The possibility of control of fungal plant diseases through specific inhibition of polyamine biosynthesis is most excited and for reaching development [16, 17].

5. Conclusion

It is expected that earth's temperature will increase by 3–5°C in the coming 50–100 years. As there is continuous increase in temperature and uneven rainfall the changes of flood and drought is always in consideration. The anthropogenic activities such as excessive fertilizers, inappropriate irrigation and exploitation of metal resources can lead to salt stress to a large extent. Under these circumstances, plants will probably encounter more frequently, concurrently both biotic and abiotic stresses. It is the duty of plant breeders to develop stress tolerant cultivars in order to secure food security and to ensure safety to the farmers. Molecular work is to be done at the genetic level to develop mechanisms in plants in order to prevent them from different types of stress conditions. Unless responsive mechanisms are not developed against biotic and abiotic stresses, the plants will continuously subjected to such stresses and ultimately will prove a great threat to world agriculture.

Author details

Audil Gull[1*], Ajaz Ahmad Lone[2] and Noor Ul Islam Wani[2]

1 Division of Genetics and Plant Breeding, SKUAST-Kashmir, India

2 SKUAST-Kashmir, India

*Address all correspondence to: audilgull26@gmail.com

References

[1] Verma S, Nizam S, Verma PK. Biotic and abiotic stress signalling in plants. Stress Signaling in Plants: Genomics and Proteomics Perspective. 2013;**1**:25-49

[2] Zhu JK. Salt and drought stress signal transduction in plants. Annual Review of Plant Biology. 2002;**53**:247-273

[3] Seki SK, Reddy KR, Li J. Abscisic acid and abiotic stress tolerance in crop plants. Frontiers in Plant Science. 2007;7:571

[4] Mahajan S, Tuteja N. Cold, salinity, and drought stress: An overview. Archives of Biochemistry and Biophysics. 2005;**444**:139-159

[5] Alcazar R, Marco F, Cuevas J, Patron M, Fernanado A, Carrasco P, et al. Involvement of polyamines in plant response to a biotic stress. Biotechnology Letters. 2006;**28**:1867-1876

[6] Bartels D, Sunkar R. Drought and salt tolerance in plants. Critical Reviews in Plant Sciences. 2005;**24**:23-58

[7] Groppa M, Benavides M. Polyamines and abiotic stress: Recent advances. Amino Acids. 2008;**34**:35-45

[8] Bagni N, Tassoni A. Biosynthesis, oxidation and conjugation of aliphatic polyamines in higher plants. Amino Acids. 2001;**20**:301-317

[9] Navakouidis E, Langebartels C, Meindl U, Kotzabasis K. Ozone impact on the photosynthetic apparatus and the protective role of polyamines. Biochem Biophys Acta. 2003;**162**:160-169

[10] Wang X, Shi G, Xu Q, Hu J. Exogenous polyamines enhance copper tolerance of Nymphoides petatum. Journal of Plant Physiology. 2007;**164**:1062-1070

[11] He L, Nada K, Tachibana S. Effects of spermidine pretreatment through the roots on growth and photosynthesis of chilled cucumber plants (*Cucumis sativus* L.). Journal of Japanese Society of Horticultural Sciences. 2002;**71**:490-498

[12] Kaur-Sawhney R, Tiburcio A, Altabella T, Galston A. Polyamines in plants. Journal of Cell and Molecular Biology. 2003;**2**:1-12

[13] Asthir B, Spoor W, Duffus C. Involvement of polyamines, diamine oxidase and polyamine oxidase in resistance of barley. Euphytica. 2004;**136**:307-312

[14] Torrigiani P, Rabiti A, Bortolotti C, Betti L, Marani F, Canova A, et al. Polyamine synthesis and accumulation in the hypersensitive response to TMV in Nicitiana tabacum. The New Phytologist. 1997;**135**:467-473

[15] Walters D. Polyamines in plant-microbe interactions. Physiological and Molecular Plant Pathology. 2000;**57**:137-146

[16] Galston A, Weinstein L. Control of phytopathogen by inhibitors of polyamine biosynthesis. Advances in Experimental Medicine and Biology. 1988;**250**:589-599

[17] Walters D. Polyamines and plant diseases. Plants Today. 1989;**22**:1

Chapter 2

The Ecophysiology of Abiotic and Biotic Stress on the Pollination and Fertilization of Cacao (*Theobroma cacao* L.; formerly Sterculiaceae family)

Puran Bridgemohan and Majeed Mohammed

Abstract

The cocoa crop growth is highly influenced by environmental conditions, viz. temperature, which influence the phenological stages of flowering, fruiting, and pod growth. The plant produces caulescent flowers that are hermaphrodite and pollinated by insects, mainly *Forcipomyia* sp. (Diptera: Ceratopogonidae), but flowers setting to pods are very low. The efficiency of pollination depends on the degree of pollen compatibility and the number of pollen grains deposited on the stigma. It is assumed that midge population can be a limiting factor in the pollination of cocoa in addition to the environmental conditions. However, populations of insect pollinators are often severely disturbed by hurricanes through flooding of essential habitat and the widespread loss of existing flowers. This chapter will explore the role of midges [biotic] and the effect of climate [abiotic] variables. Understanding these ecological dynamics can lead to ways of conserving midge populations, mitigating the effects of global climate change and extreme climatic events.

Keywords: phenological, *Forcipomyia* sp., insect pollinators, *Theobroma cacao* L.

1. Introduction

Cacao [*Theobroma cacao* L. (formerly Sterculiaceae family)] is a perennial crop in chocolate confectionary industry [1]. There are three main groups of cacao varieties, viz. Criollo (*T. cacao* var. *cacao*), Forastero (*T. cacao* var. *sphaerocarpum*), and Trinitario (hybrids of Criollo and Forastero) [2, 3]. The Criollo fruits are oblong to ovoid in shape with yellowish-white seeds. The Forastero are ellipsoid to round with a smooth surface. The Trinitario hybrid fruits are highly variable when the plants are grown by seeds. Breeding improvements have led to the "fine or flavour" cocoa beans which have high yield and quality as the Trinidad Selected Hybrids (TSH) [4].

1.1 Cacao stress

Cacao cultivation is challenged by multiple abiotic and biotic stress factors [5], as they sessile to physical environment interactions with pest and pathogens [6], evapotranspiration [7], soil salinization [8], and climate change [9]. They developed a multitude of defence mechanisms to adapt and survive stress conditions [10] and harmful microorganisms [11].

1.2 Cacao agroecology

The agroecological zones [AEZ] relate to soil fertility cycling and weed, pest, and watershed management. The cacao crop growth is specific to the AEZ, viz. temperature [12], flooding [13], and water stress [14]. The bimodal seasons influence the phenological stages of flowering, fruiting, and pod growth [15]. Water stress inhibits leaf development and pod setting and induces leaf abscission and photosynthetic rate (PR) [16].

1.3 Scope of stress pollination and fertilization

The ecophysiology of abiotic and biotic stress on the pollination and fertilization of cacao is specific to four (4) main stages in the crop reproductive cycle, viz. prepollination, pollination fertilization, and postfertilization. This treatise processes from pollen germination to ovary fusion to the young pod development or cherelle [4].

The scope of the treatise is limited to the author's research and critical review of the biotic or the internal factors that influence flower and fruit set in cacao. The abiotic factors are light intensity, relative humidity (RH), flooding, water stress, and cultural practices as shade, intercropping, and crop nutrition. The resilience of the pollinator to adapt to climatic changes and crop residue manipulation to increase the population dynamics of the cacao midges is examined, as well as advances in the area of olfactory sensitivity and cacao pollen odour [17].

2. Botany

2.1 Vegetative stage

Cacao can be propagated by seedling, cuttings, or terminal grafts, reach maturity at 12–18 months [4], and develop juvenile vertical shoot which produces lateral branches or "jorquette/chupons". It is an evergreen understory tree (20–25 m) but is shorter under cultivation (3–5 m) [18] and exhibits a flushing-type growth pattern, with 2–4 flushes/year. The fruit is a drupe but is referred to as a pod as it is fleshy and indehiscent, with internal seeds.

2.2 Cacao floral and fruit phenology

Cacao flowers prolifically (800–1000 caulescent flowers/tree) with 40 flower cushions [4]. The flowers dehisce in the afternoon and release pollen to a receptive stigma (**Figure 1**). The non-pollinated flowers abscise 24–36 h after anthesis. The flower setting to pods or cherelle is very low (0.5–5%) [19], hermaphrodite, and pollinated by insects, mainly *Forcipomyia* sp. (Diptera: Ceratopogonidae) [20]. The position of staminodes is around the style of cocoa flowers, and the stability of cocoa flowers is relative to pollination and seasonality [21]. The overall cacao pollination is low and is not significantly affected due to the small number of splay staminodes.

Figure 1.
BBCH of cacao flower from first visible observation to fully open. (a) BBCH 60: flowers open (30-31 FBV), (b) BBCH 62: 10% of flower open, (c) BBCH 65: 50% of flower open and (d) BBCH 69: 90% of flower open.

2.3 Cacao BBCH scale

The BBCH scale was used to differentiate the growth stages of cocoa steps, as it provides an insight of the floral phenology [4, 17, 22, 23]. The inflorescence emergence visible sign (Stage 5) shows flowering, that the buds primordia are 150 μm wide at the first bud visible (FBV). This continued over 30 FBV days and terminated at stage BBCH 59, when the flower bud growth is completed, but the bud is still closed. Usually, the individual flower cushion can produce many flowers at different stages of development over this growth stage (**Figure 2a–d**).

2.4 Cocoa reproductive anatomy and physiology

The cocoa flowers emerge as small cushions on the mature branches. The petal curves into a tiny hood that fits down around the style. The male flower consists of five double stamens, with each stamen consisting of up to four anthers. The female flower consists of five united carpels, each containing 4–12 locules (cavities). Due to this anatomical structure, a large insect like a honeybee in search of nectar could be a difficult venture, and only small insect as biting midge (subfamily Forcipomyiinae) would be able to pollinate the crop [4, 17]. The flower does not produce nectar, but the midges are attracted to red spikes on the flowers as the flower opens at dawn to facilitate the pollen release, and the stigma is only receptive to pollination for a period less than 12 hours. Unpollinated flowers drop off the next day with <10% successful pollination and 2% fruit development [4, 17].

Figure 2.
Cocoa cherelle wilt.

2.5 Crop ecophysiology

Cacao is a specific physiotype occupying a limited and defined ecophysiological sphere based on the crop comparative morphology and anatomy, physiology, biochemistry, biophysics, and molecular biology [24, 25]. The crop has its own set of complete phenotypical traits generated by a genotype in the morphological domain as a morphotype and in the physiological domain as a physiotype.

The crop physiotypical traits basically provide a good explanation of the occurrence of plants in their habitats and their relations between the environment and the morphological traits. The ecophysiology is influenced by both biotic and abiotic factors that affect vegetative growth and subsequent pollination and fertilization. The ecophysiology of cacao is measured by the actual behaviour of plants under natural environmental conditions, especially photosynthesis and transpiration.

2.6 Crop yield determinant

The cacao pod yield is influenced by photosynthesis and partition of photoassimilate [26]. The most important parameters for determinants of cacao yield are related to light interception and photosynthesis and photoassimilate distribution. These chains of events are modified by abiotic factors particularly during the floral phenology of the crop and influence the yield of cocoa. Yields of this crop depend upon successful transfer of pollen between flowers. Pollinator availability and efficacy can influence fruit set [27]. Recent studies have shown significant pollination limitation in several regions due to ineffective pollinators [28, 29].

3. Pollination and fertilization

Cocoa flowers are hard to pollinate due its complicated reproductive structure [29], and pollination and pod set depend on *the* degree of pollen compatibility, quality [30], and germination [31]. Pollination is the result from the visit of the single pollinator [32], and fertilization is influenced by a combination of plant morphological traits (biotic) and climatic variables (abiotic).

3.1 Cacao pollination

The main pollinators are midges (Diptera: Ceratopogonidae) and other small Diptera (Cecidomyiidae) [33–35]. The flowers are minute, so only a few taxa can be effective pollinators (**Figure 3**). The Ceratopogonidae are effective pollen carriers as

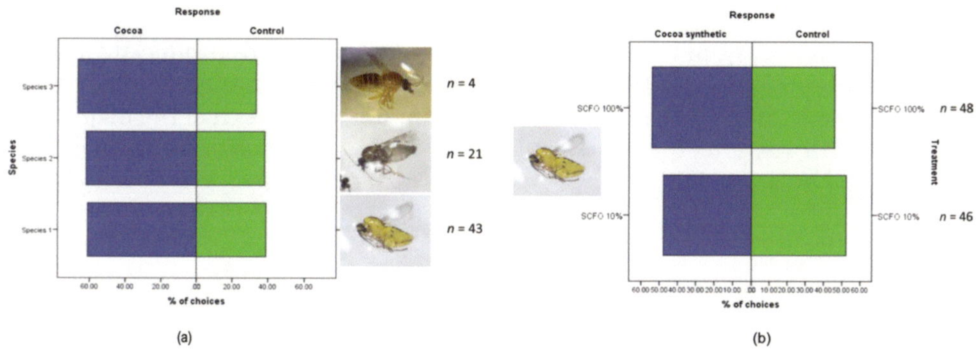

Figure 3.
(a) Response of midges to natural cocoa flower odour and (b) response of midges to synthetic blend mimicking cocoa flower odour.

the flower produces no nectar for the midges to collect, and it is suggested that they may be attracted to an odour or pheromone [36].

3.2 Pollinator dynamics

Midge population are threatened by tropical hurricanes and flooding. These small poor-flight insects are easily swept away by high winds [4, 17]. Midges would normally thrive in moist humid environment [37, 38], but excessive rain or drought could decimate the natural population. Bridgemohan et al. [4] examined the relationship between the midge population, flower pollination, and selected weather variables in several different Caribbean cocoa-producing islands. They recommended the manipulation of cocoa and banana residues as habitat for the adult midges to complete their reproductive cycle. This has improved crop yield through improved pollination and enhanced fertilization (**Table 1**).

3.3 Pollination intensity

There is a relationship between pollination intensity, fruit survival, and cocoa seed production [39]. The pollen/seed ratio increased with increased pollination intensities and seed yield.

3.4 Pollen germination and fertilization

Cacao genotypes are self-compatible and self-incompatible [19]. The pollen germination and fruit set will not occur after self-pollination of an incompatible

Months	Cacao leaf litter	Cocoa pods	Banana pseudostems	x̄ [SE]
March	4.75	4.25	4.75	4.6±[0.17]
April	5	4.25	2.75	4.0±[0.66]
May	3.5	3.25	1.5	2.8±[0.63]
June	2.75	1.75	2	2.2±[0.30]
July	5	8.25	7	6.8±[0.95]
August	11.5	9.5	11.75	10.9±[0.71]
x̄ [SE]	5.4 ±[1.27]	5.2±[1.23]	5.0±[1.59]	

Table 1.
Population of midges harvested from the Centeno [Trinidad] location.

genotype. Pollen germination improves (95%) with enclosed pollination due accumulation of CO_2, but fruit set can be low (45%) and seeds produced from self-pollinations are high (95%) compared to the cross-pollination (100%). The ovary 48 h after showed that the self-pollination varieties had the majority of ovules with a zygote, but some ovules contained unfused male and female gametes and polar nuclei.

3.5 Compatibility

Compatibility is highly variable in cacao. Self-incompatibility is higher in Criollo and Trinitario types, but Forastero cultivar is self-compatible with a high degree of cross-pollination [40]. Pods in self-compatible trees (50%) are cross-fertilized. It was inconclusive that all pods were the result of cross-fertilization. The fertilization of different flowers by different male parents or by mixed pollen was evident.

3.6 Climatic effects

The impact on pollinator distribution and abundance in tropical agricultural system areas are more likely to be affected by climate change with reduced crop yields [41, 42]. When rainfall and humidify cannot be managed, it will interfere with the efficacy of pollination and fertilization [37].

4. Effects of abiotic factors

Abiotic stresses affect the internal metabolic processes of plant and reduce their efficiency in dry matter production, accumulation, and partitioning [43]. There is an interaction between climatic variables, crop management, cacao pollination, and adult midge pollinators [37].

4.1 Light intensity

Cacao is limited by light interception and photosynthesis due to external and internal shade (self-shading) and light extinction inside the canopy [44, 45]. Heavy shade reduces seed yield because of low photosynthate production and partitioning [46, 47] and increases the incidence of diseases. Cacao is a shade-tolerant plant, and appropriate levels of shading could improve photosynthesis and seed yield [4] and reduce excessive evapotranspiration [48] and can tolerate decreased humidity and high temperature stresses during the dry season [21].

4.2 Relative humidity (RH)

Humidity is predictor of insect abundance, but it is difficult to predict its impact on pollination in cocoa [37]. Stomatal opening is related to relative humidity [49] but does not show high stomatal resistance under water stress or low RH [50]. Some genotypes are more sensitive to low RH which can be a limiting growth factor [49] and result in a reduction of net photosynthesis and low water-use efficiency [46, 50].

4.3 Rainfall

The stability of the cocoa flowers depends on the season and % pollination [51]. In the dry season, unpollinated flowers showed a low flower stability (72%)

compared to the wet season (94%). This improvement indicated that seasonal factors (water stress) have drastic effects on yields. The crop requires a high rainfall (1200–2500 mm/yr), and the rainfall over the previous month is a significant predictor of cocoa midge abundance, as the larvae require moist, decomposing vegetation (cocoa pods and leaf litter) that is more abundant in the wet season [52, 53].

4.4 Flooding

The tropics have a distinct dry and a wet season with hurricanes, excessive rainfall, and flooding [4, 47]. Flooding affects the initial crop growth when the total rainfall exceeds evapotranspiration. The hypoxic conditions in the hydromorphic soil [13] result in decreased vegetative growth, photosynthesis, flowering, and pod production [13, 24, 54, 55].

4.5 Water stress

Almeida et al. [14] found that drought resistance occurred through osmotic adjustment in most cacao genotypes and that many maintained relative water content (90%) and leaf water potential (Y_w = −1.0 MPa) are gradually decreasing (55% at −3.5 MPa). They found that there was a significant increase in leaf [K and P] during the dehydration process of some genotypes (Y_w = −1.5 MPa). Water stress affects internal water availability, translocation of assimilates, sink-source relationship, and flower set and cacao bean development [45, 47].

4.6 Shade

As a shade crop, cacao bean yields are considerably low due to shade-intercrop competition for water, nutrients, and light [56]. Traditionally, it was assumed that shade was critical regardless of yield [46, 57]. However, the cacao can tolerate full sunlight and produce more pods than under shade [58]. Under fundamental cultural practices, increased in crop density, aeration, and sunlight penetration seed production are enhanced [59, 60]. Moderate shade hardly affects bean yield, compared to heavy shading (>60%) which reduces yields by 30% [61].

4.7 Intercropping

Intercropping manipulates the agroecological conditions and enhances yield by promoting effective mutualism between species. In pure stand/shade tree-cocoa systems, there is inconclusive evidence on species interactions and competition under two separate shade species (*Terminalia superba* and *Newbouldia laevis*) [62]. Intercropping had no effect on cocoa biomass production in comparison to monoculture cocoa. Shading induced foliage and root formation both with and without fertilization. Light affects growth in the absence of underground competition. Intercropping can suppress K uptake in cocoa foliage (25%), due to interspecific competition.

5. Effects of biotic factors

Biotic stress affects pathogens, and there is an interaction between abiotic and biotic factor stresses (osmotic, ionic, or exogenous) [63, 64]. These biotic stress factors can affect pollination and fertilization through its physiology, nutrition, pest and disease, hormones, and pollinators and predators.

5.1 Physiological

Environmental factors affect cacao growth (morphology and physiology) in the orthotropic phase especially light intensity, photosynthetic capacity, and chlorophyll content [48, 63, 65–67]. Shaded leaves exhibit greater variability than full sunlight [68]. Leaf blade thickens with the increased light intensity regardless of NO_3 concentration, but under shade with high [N] there is a reduction [69]. Under full sunlight, rates of leaf expansion are low due to excessive transpiration and fewer stomata per unit leaf [58]. Cacao exhibits increased rates of net photosynthesis (400–750 $\mu mol\ m^{-2}\ s^{-1}$), which reflects 20–30% of PAR at full sunlight, compared to shaded (3–4%) [70]. Once the nutritional demands of the crop are met, the yield is dependent mainly on solar radiation [71]. Partial pruning of cacao reduces the fine root production resulting in a reduction photoassimilation and internal competition between vegetative flushing and root formation [72].

5.2 Crop nutrition

Mineral deficiencies or surpluses can lead to poor growth and development or toxicity to physiological processes, inclusive of pollination and fertilization. In cocoa under shade, N, P, and K uptake can be increased by 54, 112, and 71%, respectively. Intercropping with shade trees may not increase cocoa biomass; however, nutrient uptake is sustained for N and P due to low interspecific competition [62]. The crop utilizes 700 kg K/ha to produce 1000 kg/year of seeds including biomass. There is an inverse relationship between leaf transpiration and K doses [72]. Cacao trees well supplied with K are more tolerant to adverse water stress [73]. Regardless of the source of K, it induced low stomatal conductance and transpiration rate and improved WUE but without affecting shoot dry biomass [73].

5.3 Plant growth regulators

Plant hormones in cacao facilitate the adaptation to changing environments through mediating nutrient allocation and source/sink transitions [17]. ABA controls abscission, and ethylene accelerates it in the presence of ABA, but naphthalene acetic can prevent it without inducing fruit set. Cytokinins and auxins effects could be either synergetic or antagonistic depending on the interactions to other abiotic stresses. Unpollinated cacao flowers (90%) abscised immediately after anthesis due to increased ABA levels [19]. Fluridone inhibits ABA production, abscission zone layer, and senescence of the flowers, while ethylene production increases only after anthesis. Aminoethoxyvinylglycine (AVG) application can delay abscission compared to NAA + AVG.

5.4 Pollinators and predators

The key pollinator of cacao is the midge (Diptera: *Ceratopogonidae*) *in* which the plant receives self or outcross pollen and/or exports pollen to a conspecific plant [74]. This can be interfered by disturbances to the insect natural habitat and use pesticide [4, 47], resulting in low fruit set [75]. Conservation of the pollinator improves crop yield by removing practices that alleviate the negative impacts on the insect's habitat [37]. Discarded cacao pod increased the number of fruits/tree (**Table** 1) indicating a promotion of the pollinator in the ecosystem [4, 17]. Cacao pod residue increases the population of the two predators to the midges, viz. spiders and skinks, but these predators do not inhibit pollination. Improvements to midge habitat increased the availability of alternative habitat and food resources for both the pollinators [38]. Biological conservation and good agricultural practices improved pollinator efficacy and species conservation [75].

5.5 Diseases

Cacao is adapted to areas with high humidity climates but is pre-disposed to the risk of fungal diseases. In some areas, the cocoa crop is affected by various diseases (30–40%) which attack their vegetative parts and fruits. The crop is most susceptible at the fruiting cycle regardless of the genotype [76, 77]. Adomako [78, 79] found that cocoa experienced high yield loss due to unusable pods (10%), black pods (64%), immature ripe pods (30%), rodent damage (4%), and other damages (1.3%) which is close to 98 $kg\ ha^{-1}\ yr^{-1}$ of dry beans. The four major diseases are witches' broom disease, frosty pod rot, *Phytophthora* pod rot, and vascular-streak dieback (VSD).

Witches' broom disease (*Moniliophthora perniciosa*) attacks growing tissue causing cocoa trees to produce branches with no fruit and ineffective leaves. The epidemiology is synchronized with the crop phenology, and the spread and reproduction of the fungus depend on the availability of water. Basidiospores are released at night with high humidity (80%) and temperatures (20–30°C) but have a short viability period and are sensitive to light and drying. The pathogen is also spread by infected seeds or budwood. Host resistance is the best option for control, such as the Trinidad selected hybrids, and treatments of fungicides and phytosanitary pruning have proven to be effective [77, 80, 81].

Frosty pod rot (basidiomycete: *Moniliophthora roreri*) infects young pods (1–3 months) with a white fungal mat on the pod surface. The dry powdery form of spores can be dislodged by water, wind, or physical disturbance of the pod and is spread easily. Disease incidence varies with the cultivar, pod age, and high rainfall. Application of systemic copper fungicides is essential when the crop is to be propagated by seeds or budwood. All cocoa species are susceptible to this disease, but fungicides (Flutolanil) and quarantine of infected fields can be effective [77, 80, 81].

Pod rot or black pod is caused by the fungus *Phytophthora* spp. (*P. palmivora*, *P. megakarya*, and *P. capsici*.) resulting in high yield loss (20–30%) and tree deaths (10%) (**Figure 4**). *P. megakarya* is the most important pathogen in cocoa in Africa and *P. capsici* in Central and South America, causing significant rotting or necrosis of pod losses especially in favourable environments [77, 80, 81]. Infestation can occur at any stage of development with the initial symptoms appearing as small, hard, and dark spots on the pod. Internal tissues and the beans are colonized and leaving a shrivelled pod.

Pod infected with *P. palmivora* produces up to 4 million sporangia which are disseminated by rain, ants, flying insects, rodents, and bats, including contaminated pruned branches. *P. megakarya* sporulation is usually more abundant with a soil borne phase which causes root infection and maintains a reservoir of inoculum in the soil surface water.

Treatment with systemic copper fungicides (metalaxyl) is frequently recommended together with injections of inorganic salt and potassium phosphonate,

Figure 4.
(a) Banana psuedostem as used as breeding site for midges and (b) cacao pod used as breeding site for midges.

which have proven to be effective against *P. palmivora*. Farm management practices which optimize shade and aeration can reduce surface wetness effectively. The complete harvesting, sanitation, and disposal of infected pods and husks can reduce the disease [77, 80, 81].

This disease is caused by *Oncobasidium theobromae* with its characteristic symptoms of chlorosis of the leaf. The fungus spreads internally within the plant resulting in death. Infected leaf litter in the rainy season is the main source of basidiospore discharge and spreads mainly at night and by wind or high humidity [77, 80, 81]. The spores have short lifespan. Protective fungicides, open canopy, and control of shading which increases aeration and insolation of the foliage are effective.

5.6 Insect pests

Insect pests are less destructive to the cacao floral and reproductive organs compared to rodents, birds/parrots, and monkeys. They are classed in three groups, viz. cause primary damage, transmit disease, or rise to pest status due to tolerance to insecticides. The two major pests are the mirids and the cocoa pod borer (CPB).

Mirids (capsids) are the major cacao insect pests (*Distantiella theobroma*, *Sahlbergella singulari*, *Helopeltis* spp., and Monalonion spp.) which reduce yields (75%) by feeding on the stem, shoots, and pods producing necrotic lesion causing dieback. Female mirids lay up to 60 eggs inside the pod husk, thus spoiling the beans. The insect is more attracted to trees in full sunlight but feeds and inhabits on the shady areas of trees [27, 82].

The organochlorine insecticides and Imidacloprid (Actellic/Talstar and Promecarb) are effective, but reduced insecticide is recommended to allow the natural enemies to increase for biological control. Integrated pest management (IPM) can control mirid using black ant (*Dolichoderus thoracicus*) and weaver ant (*Oecophylla smaragdina*).

The cocoa pod borer (*Conopomorpha cramerella*) causes losses in both young and mature cocoa pods. The main symptom of infested pods is premature ripening resulting in poor bean quality. Contact pyrethroid or carbamate insecticides on the undersides of lower branches can keep economic damage levels to a minimum. The fungus *Beauveria bassiana* can infect larvae and pupae. Traps with synthetic pheromones or female pod borer moths are used to reduce male's population.

5.7 Olfactory sensitivity and cacao pollen odour

The cocoa flowers appear to have no discernible odour like citrus, and it is only the pollen and nectar that are the pollinator's attractants. However, Erickson et al. [83] found floral fragrance in cultivated *T. cacao* consisted of 78 components which are mainly saturated and unsaturated hydrocarbons, with 1-pentadecene or *n*–pentadecane. Arnold et al. [37] *found that* three species of cocoa midge were attracted to the natural odours of cocoa flowers. *Dasyhelea* cf. *borgmeieri* was not attracted to a synthetic cocoa flower odour suggesting that it is the minor component of the cocoa flower's odour that attracts midges.

6. Cherelle wilt and fruit abortion

Young pods or cherelles are lost to physiological thinning known as cherelle wilt [84, 85]. Many of the cherelles die later with cherelle wilt as a natural event or become infected by fungi or bacteria. The first cherelle wilt occurs at 7 weeks after pollination (WAP) with a second wilt later (10 WAP) and has larger embryos and smaller pod stalk pods. Young fruit abortion is high (10%) with a reduction in bean weight (3–10%) [86]. Pollination and assimilate limitation cause low fruit: flower ratios in cacao [27]. The "wilting phase" exhibits many changes in the anatomical structure including swelling of the pod, enlarged vascular, and lignification of the middle pericarp [85]. The xylem of the fruit of cherelle wilt was caused by occlusions in the xylem vessels of the fruit-stalk, which is associated with the oxidative activity of a cambial or meristematic tissue [86].

6.1 Biotic effects on cherelle wilt and fruit abortion

It is postulated that both types of wilt arise as a result of biotic and abiotic factors. Lachenaud [87] examined the stages between pollen germination and ovule fusion and found incomplete pod filling occurred mainly after physiological heritage wilt. The amount of fallen flowers with set ovary is insignificant (0.5%), suggesting that flowers pollinated with sparse pollen grains fell without setting. When fruit setting occurred at the same time on the same trees, wilted cherelles contained significantly more fertilized ovules than pods with beans.

The lack of hormones produced by the endosperm causes a decrease in the uptake of water and minerals, thereby inducing wilt. Wilted pods contained less cytokinin-like substances than healthy pods [88]. Auxin accumulates in deficient tissues may be responsible for incomplete filling and parthenocarpy [84]. Wilt is associated with increased levels of tricarboxylic acid cycle intermediaries and decreased levels of major metabolites in the biosynthetic pathways and regulation of abscisic acid and cytokinin levels [24].

There is inverse relationship between the wilt index and endogenous growth compounds in cacao [85, 86], with more polyphenol oxidase activity in the inner and outer pericarps of pods [84]. The pericarp and seed development are largely independent processes except for the inception of fruit growth and the changeover from the wilting to the non-wilting phase.

Despite the abundant flowering in cacao, a small number of cacao flowers (0.5–5%) become pollinated, and others become cherelles. Further to the low pod set of trees, few cherelles develop into mature pods with up to 75% of cherelles lost to the thinning condition. This may be due to lower level of assimilates available to the cherelle due to severe intra-plant competition [38, 89] and inefficient partitioning of photoassimilates [4]. Cherelles are attacked by insect (*stramenopile*) and fungi (*Phytophthora* and *perniciosa*) in early stages and frosty pod rot (*Moniliophthora roreri*), and rodents, squirrels, and parrots are common in mature pods. Pollination efficiency is negatively correlated with the number of flowers, and cherelles produced also varied between and within crosses [90]. Higher yielding trees were more efficient in converting flowers into pods mainly due to events that occur at the late cherelle stage.

6.2 Abiotic effects on cherelle wilt and fruit abortion

There are several environmental or abiotic factors that induce cherelle wilt, but these may have an interaction with the crop physiology. The cocoa tree allows as many cherelles to develop into mature pods based on nutrient availability, but those that do not fall become mummified and decayed [4]. There are significantly higher levels of nutrients (N and P) in soils under shaded than in un-shaded, corresponding with lower populations of wilts. Incomplete pod filling seems to be due to interactions between nutritional factors and genotypes [91].

Cherelle wilt is higher in un-shaded crop due to moisture stress, higher evapotranspiration, and lower nutrient available to support crop yield. There was a little difference between the effects of the wet and dry conditions on flower production or setting or on cherelle wilt [92]. Cacao planted during the dry period developed few flowers, but initiation was apparently stimulated; in a subsequent wet or medium period, flowering was exceptionally heavy, but pod setting was poor and cherelle wilt was high [93].

Shaded cocoa has lower light intensity and interception but enhanced nutrient cycling and improved healthy pod development [4]. The level of overhead shade provided by the forest significantly influences litter fall, decomposition, and soil fertility and development of cocoa pods. Under un-shaded farms, litter fall is very high, but the rates of litter decomposition are very slow. Cacao fruit losses increase due to physiological wilt associated with higher temperatures, but differed between genotypes, reflecting genetic variation and also competition for assimilates between vegetative and reproductive stages [65].

6.3 Cherelle management

Cherelle wilt can be reduced by improving the health of cocoa trees through the *application of Fertilizers and mulches* and sunscald control. Ethephon application to the pedicel of young fruit resulted in morphological changes similar to natural wilt, suggesting that cherelle wilt could be reduced with indole butyric acid and gibberellic acid [94]. Cherelle wilt will increase due to high crop density and competition for nutrients, water, and light. Moist but well-drained soils and mulches will reduce cherelle. There is a balance between biotic and abiotic factors that determine the optimum number of cherelles that a tree can sustain, which is dictated by the demands of the "sinks" or developing beans within the surviving pods [4].

7. Conclusion

The interaction of environmental and plant genetic characteristics dictate the survival and reproductive efficacy of the cacao. Although the plant can produce its optimum yield of flowers, pollination and fertilization efficiency are under the influence of the equilibrium of the biotic and abiotic variables and their mutualistic interactions. The plant can only sustain a certain number of young pods or cherelles to full maturity. However, internally the number of beans/pods is still subjected to the partitioning of photoassimilates to ensure optimum bean filling. Regardless of the efficiency of pollination and fertilization, intra-plant competition for photo-assimilates will result in a high number of beans/pod that will be incomplete or poorly filled and reduced bean weight and final yield.

The ecophysiology of pollination and fertilization of cacao beans is manageable by optimized agrocultural practices. Adequate midge breeding sites using cocoa pod and banana pseudostem can improve the insect population and subsequently increase pollination and fertilization [95]. This management practice is envisaged as the way to increase bean yield. The cocoa flowers are influenced by seasonality, weather conditions (abiotic) and pollination (biotic). The dynamics of cocoa pollination involves harmonization pollinator population cycle and the flower phenology. However, the numbers of cocoa-pollinating midges are lower in the dry season but increase in the wet season, but the natural habitat is pre-disposed to flooding, and insect flight is curtailed by high winds and rainfall.

Cocoa leaf litter and the pod husk constitute the bulk of ground material but dry up and become unsuitable in the dry season. Also, flower cushion is affected by water stress and high relative humidity and encourages flower abortion. In the favourable wet season, there could still be reduced pollination due to less pollinators. The crop self manages it photoassimilate and its partitioning to avoid significant intra-plant competition, and to obtain functional balance, significant amount of flowers are abscised. A better understanding of the biotic and abiotic variables of pollination and fertilization processes and the midge biology and ecology has led to the development and validation of manipulation of the insect natural breeding site with an increase of insect population dynamics. This was amply demonstrated with increased pod yield and number of beans per pod.

Author details

Puran Bridgemohan[1]* and Majeed Mohammed[2]

1 Crop Science, Biosciences Agriculture and Food Technology, University of Trinidad and Tobago, Trinidad and Tobago

2 Post Harvest Physiology, University of West Indies, Trinidad

*Address all correspondence to: puran.bridgemohan@utt.edu.tt

References

[1] Alverson WS, Whitlock BA, Nyffeler R, Bayer C, Baum DA. Phylogeny of the core Malvales: Evidence from sequence data. American Journal of Botany. 1999;**86**:1474-1486

[2] Bartley BGD. The Genetic Diversity of Cacao and Its Utilization. Wallingford, UK: CABI Publishing; 2005. 341 pp

[3] Motamayor JC. Identification of marker-trait associations for self-compatibility in a segregating mapping population of *Theobroma cacao* L. Tree Genetics & Genomes. 2011;**7**(6):1159-1168

[4] Bridgemohan P, Mohamed MES, Mohammed M, Singh K, Bridgemohan RSH. The application of BBCH scale for codification and illustrations of the floral stages of Caribbean Fine cacao *Theobroma cacao* L. Journal of Agricultural Science and Technology. 2016:1-10. DOI: 10.17265/2161-6256/2016.01.001

[5] Kissoudis C, van de Wiel C, Visser RG, van der Linden G. Enhancing crop resilience to combined abiotic and biotic stress through the dissection of physiological and molecular crosstalk. Frontiers in Plant Science. 2014;**5**:207. DOI: 10.3389/fpls.2014.00207

[6] Peters GP, Marland G, Le Quéré C, Boden T, Canadell JG, Raupach MR. Rapid growth of CO_2 emission after the 2008-2009 global financial crisis. Nature Climate Change. 2011;**2**: 1-3. DOI: 10.1038/nclimate1332

[7] Zhao M, Running SW. Drought-induced reduction in global terrestrial net primary production from 2000 through 2009. Science. 2010;**329**(5994):940-943. DOI: 10.1126/science.1192666

[8] Munns R, Tester M. Mechanisms of salinity tolerance. Annual Review of Plant Biology. 2008;**59**:651-681. DOI: 10.1146/annurev.arplant.59.032607.092911

[9] Garrett KA, Dendy SP, Frank EE, Rouse MN, Travers SE. Climate change effects on plant disease: Genomes to ecosystems. Annual Review of Phytopathology. 2006;**44**:489-509. DOI: 10.1146/annurev.phyto.44.070505.143420

[10] Pieterse CM, Leon-Reyes A, Van Der Ent S, Van Wees SC. Networking by small-, molecule hormones in plant immunity. Nature Chemical Biology. 2009;**5**:308-316. DOI: 10.1038/nchembio.164

[11] Grover M, Ali SZ, Sandhya V, Rasul A, Venkateswarlu B. Role of microorganisms in adaptation of agriculture crops to abiotic stresses. World Journal of Microbiology and Biotechnology. 2011;**27**(5):1231-1240

[12] Daymond AJ, Hadley P. The effects of temperature and light integral on early vegetative growth and chloroplyll fluorescence of four contrasting genotypes of cacao (*Theobroma cacao*). Annals of Applied Biology. 2004;**145**:257-262

[13] Sena Gomes AR, Kozlowski TT. The effects of flooding on water relations and growth of *Theobroma cacao* var. catongo seedlings. Journal of Horticultural Science and Biotechnology. 1986;**61**:265-276

[14] Almeida A-AF, Brito RCT, Aguilar MAG, Valle RR. Water relations' aspects of *Theobroma cacao* L. clones. Agrotrópica. 2002;**14**:35-44

[15] Cazorla IM, Aidar T, Milde LCE. Perfis do lançamento foliar, da floração, da bilração e de estágios do fruto do cacaueiro no Estado da Bahia, no período de 1987/1988. In: Boletim Técnico, Ceplac/Cepec, Ilhéus, Brasil, 1988. 58 p

[16] Hutcheon WV. Physiological studies on cocoa (*Theobroma cacao*) in Ghana (doctoral dissertation. University of Aberdeen). Implications for a Theobroma pollination syndrome. Journal of Chemical Ecology. 1981;**20**:2687-2703

[17] Bridgemohan P, Singh K, Cazoe E, Perry G, Mohammed A, Bridgemohan RSH. Cocoa floral phenology and pollination: Implications for productivity in Caribbean Islands. Journal of Plant Breeding and Crop Science. 2017;**9**(7):106-117

[18] Hosseini BS, Trueman SJ, Nevenimo T, Hannet G, Bapiwai P, Poienou, et al. Effects of shade-tree species and spacing on soil and leaf nutrient concentrations in cocoa plantations at 8 years after establishment. Agriculture, Ecosystems & Environment. 2017:134-143

[19] Aneja M, Gianfagna T, Ng E. The roles of abscisic acid and ethylene in the abscission and senescence of cocoa flowers. Plant Growth Regulation. 1999;**27**(3):149-155

[20] Dias LAS, Kageyama PY, Castro GCT. Divergência genética multivariada na preservação de germoplasma de cacau (*Theobroma cacao* L.). Agrotrópica. 1997;**9**:29-40

[21] Frimpong EA, Gemmill-Herren B, Gordon I, Kwapong PK. Dynamics of insect pollinators as influenced by cocoa production systems in Ghana. Journal of Pollination Ecology. 2011;**5**:74-80

[22] Lehrian DW, Keeney PG. Changes in the lipid components of seeds during growth and ripening of cacao fruit. Journal of the American Oil Chemists' Society. 1980;**57**:61-65

[23] Niemenak N, Cilas C, Rohsius C, Bleiholder H, Meier U, Lieberei R. Phenological growth stages of cacao plants (*Theobroma* sp.): Codification and description according to the BBCH scale. Annals of Applied Biology. 2010;**156**(1):13-24

[24] Bertolde FZ, Almeida AAF, Pirovani CP, Gomes FP, Ahnert D, Baligar VC, et al. Physiological and biochemical responses of *Theobroma cacao* L. genotypes to flooding. Photosynthetica. 2012;**50**(3):447-457

[25] Melnick RL, Strem MD, Crozier J, Sicher RC, Bailey BA. Molecular and metabolic changes of cherelle wilt of cacao and its effect on *Moniliophthora roreri*. Physiological and Molecular Plant Pathology. 2013;**84**:153-162

[26] Sounigo O, Lachenaud P, Bastide P, Cilas C, N'Goran J, Lanaud C. Assessment of the value of doubled haploids as progenitors in cocoa (*Theobroma cacao* L.) breeding. Journal of Applied Genetics. 2003;**44**:339-353

[27] Bos MM, Steffan-Dewenter I, Tscharntke T. Shade tree management affects fruit abortion, insect pests and pathogens of cacao. Agriculture, Ecosystems & Environment. 2007;**120**(2-4):201-205

[28] Groeneveld JH, Tscharntke T, Moser G, Clough Y. Experimental evidence for stronger cacao yield limitation by pollination than by plant resources. Perspectives in Plant Ecology, Evolution and Systematics. 2010;**12**(3):183-191

[29] Milius S. The flowers that give us chocolate are ridiculously hard to pollinate. Magazine. 2018;**193**(4):4 Available from: https://www.sciencenews.org/article/flowers-give-us-chocolate-are-ridiculously-hard-pollinate

[30] Lanaud C, Sounigo O, Amefia YK, Paulin D, Lachenaud P, Clement D. Nouvelles données sur le fonctionnement du système d'incompatibilité du cacoyer et ses conséquences pour la sélection. Café Cacao Thé. 1987;**31**:267-277

[31] Hasenstein KH, Zavada M. Auxin modification of the incompatibility response in *Theobroma cacao*. Physiologia Plantarum. 2001;**112**:113-118

[32] Yamada MM, Guries RP. Mating system analysis in cacao (*Theobroma cacao* L.). Agrotrópica. 1998;**10**:165-176

[33] Billes DJ. Pollination of *Theobroma cacao* L. Tropical Agriculture (Trinidad). 1941;**17**:151

[34] Posnette A. Pollination of cacao in Trinidad. Tropical Agriculture (Trinidad). 1994;**21**:115-118

[35] Salazar-Díaz R, Torres-Coto V. Estudio de la dinámica de polinizadores del cultivo de cacao (*Theobroma cacao*) en tres sistemas de producción. Revista Tecnología en Marcha. 2017;**30**:90-100

[36] Anon. Optimisation of cocoa pollination for increased cocoa yields and income generation. Seminar. April 2016

[37] Arnold SEJ, Bridgemohan P, Perry GB, Spinelli GR, Pierre B, Murray F, et al. The significance of climate in the pollinator dynamics of a tropical agroforestry system. Agriculture, Ecosystems & Environment. 2018;**254**:1-9

[38] Bridgemohan P, Bridgemohan RSH. Crop nutrition studies on grain filling and chalkiness in rice. Journal of Plant Breeding and Crop Science. 2014;**6**(10):144-152

[39] Falque M, Vincent A, Vaissiere BE, Eskes AB. Effect of pollination intensity on fruit and seed set in cacao (*Theobroma cacao* L.). Sexual Plant Reproduction. 1995;**8**(6):354-360

[40] Harland SC, Frecheville GE. Natural crossings and the genetics of axil spot in cacao. Genética. 1927;**9**:279-288

[41] Vanbergen AJ. The Insect Pollinators Initiative. Threats to an ecosystem service: Pressures on pollinators. Frontiers in Ecology and the Environment. 2013;**11**:251-259

[42] Josef S, Bishop J, Potts SG. Climate change impacts on pollination. Nature Plants. 2016;**2**:7

[43] Gill, Singh S, Tuteja N. Reactive oxygen species and antioxidant machinery in abiotic stress tolerance in crop plants. Plant Physiology and Biochemistry. 2010;**48**(12):909-930

[44] Bridgemohan P. Production and partitioning of dry matter in leren (*Calathea allouia* (Aubl.) Lindl). Journal of Agriculture of the University of Puerto Rico. 2011;**95**(1-2):35-43

[45] Bridgemohan P, Bridgemohan RSH. Effect of initial stem nodal cutting strength on dry matter production and accumulation in cassava (*Manihot esculenta* Crantz). Journal of Plant Breeding and Crop Science. 2014;**6**(6):64-72

[46] Alvim PT. Cacao. In: Alvim PT, Kozlowski TT, editors. Ecophysiology of Tropical Crops. London: Academic Press; 1977. pp. 279-313

[47] Bridgemohan P, Bridgemohan RSH. | Evaluation of anti-lodging plant growth regulators on the growth and development of rice (*Oryza sativa*). Journal of Cereal and Oilseeds. 2014;**5**(3):12-16

[48] Müller MW, Serrano-Minar P, Biehl B. Photosynthetic characteristics during development of leaves from *Theobroma cacao* L. Physiologia Plantarum. 1992;**853**:A105, 599

[49] Sena Gomes AR, Kozlowski TT, Reich PB. Some physiological responses of *Theobroma cacao* var. catongo seedlings to air humidity. New Phytologist. 1987;**107**:591-602

[50] Wood GAR. Environment. In: Wood GAR, Lass RA, editors. Cocoa. London: Longman; 1985. pp. 38-78

[51] Frimpong EA, Gordon I, Kwapong PK, Gemmill-Herren B. Dynamics of cocoa pollination: tools and applications for surveying and monitoring cocoa pollinators. International Journal of Tropical Insect Science. 2009;**29**(2):62-69

[52] Winder JA, Silva P. Cacao pollination: Microdiptera of cacao plantations and some of their breeding places. Bulletin of Entomological Research. 1972;**61**:651-655

[53] Adjaloo M, Banful BKB, Oduro W. Evaluation of breeding substrates for cocoa pollinator, *Forcipomyia* spp. and subsequent implications for yield in a tropical cocoa production system. American Journal of Plant Sciences. 2013;**4**:204-211

[54] Rehem BC. Respostas fisiológicas de clones de *Theobroma cacao* L. ao alagamento do substrato [M.Sc. dissertation]. Ilhéus: Universidade Estadual de Santa Cruz; 2006

[55] Mariano AH, Monteiro WR. Melhoramento genético do cacau – avaliação preliminar de cultivares, ensaio 30. In: Informe Técnico, Comissão Executiva do Plano da Lavoura Cacaueira. Ilhéus, Brazil: Centro de Pesquisas do Cacau; 1982. pp. 5-7

[56] Cheesman EE. The economic botany of cacao. A critical survey of the literature to the end of 1938. Tropical Agriculture. 1932;**9**(6):1, 16

[57] Evans H, Murray DB. A shade and fertilizer experiment on young cocoa. In: Annual Rep. Cocoa Res., Imp. Coll. Trop. Agric., British Caribbean; 1953. pp. 67-76

[58] Okali DUU, Owusu JK. Growth analysis and photosynthetic rates of cocoa (*Theobroma cacao* L.) seedlings in relation to varying shade and nutrient regimes. Ghana Journal of Agricultural Science. 1975;**8**:51-67

[59] Vernon AJ, Sundaram S. Current cocoa research. In: Proc. 4th Int. Cocoa Res. Conf., St. Augustine, Trinidad and Tobago. 1972. pp. 689-693

[60] Ahenkorah Y, Halm BJ, Appiah MR, Akrofi GS, Yirenkyi JEK. Twenty years' results from a shade and fertilizer trial on Amazon cocoa (*Theobroma cacao*) in Ghana. Experimental Agriculture. 1987;**23**:31-39

[61] Zuidema PA, Leffelaar PA, Gerritsma W, Mommer L, Anten NP. A physiological production model for cocoa (*Theobroma cacao*): Model presentation, validation and application. Agricultural Systems. 2005;**84**(2):195-225

[62] Isaac ME, Ulzen-Appiah F, Timmer VR, Quashie-Sam SJ. Early growth and nutritional response to resource competition in cocoa-shade intercropped systems. Plant and Soil. 2007;**298**(1):243-254

[63] Goellner K, Conrath U. Priming: It's all the world to induced disease resistance. European Journal of Plant Pathology. 2008;**121**:233-242. DOI: 10.1007/s10658-007-9251

[64] Wang ZY, Xiong L, Li W, Zhu JK, Zhu J. The plant cuticle is required for osmotic stress regulation of abscisic acid biosynthesis and osmotic stress tolerance in Arabidopsis. The Plant Cell. 2011;**1**:tpc-110

[65] Daymond AJ, Hadley P. Differential effects of temperature on fruit development and bean quality of contrasting genotypes of cacao (*Theobroma cacao*). Annals of Applied Biology. 2008;**153**(2):175-185

[66] Batista LP, Alvim R. Efeitos da intensidade luminosa e do genótipo sobre o crescimento em altura do fuste do cacaueiro. Revista Theobroma. 1981;**11**:61-76

[67] Abo-Hamed S, Collin HA, Hardwick K. Biochemical and physiological

aspects of leaf development in cocoa (*Theobroma cacao* L.). VII. Growth, orientation, surface structure and water loss from developing flush leaves. New Phytologist. 1983;**95**:9-17

[68] Merkel U, Müller MW, Serrano-Minar P, Biehl B. Light intensity influence on the characteristics of the photosynthetic apparatus from cocoa tree (*Theobroma cacao* L.) during leaf development. In: Proceedings of the 11th International Cocoa Res. Conf., Yamoussoukro, Côte D'Ivoire. 1994. pp. 645-653

[69] Costa LCB, Almeida A-AF, Valle RR. Crescimento, teor de clorofila e estrutura anatômica em plântulas de Theobroma cacao submetidas a diferentes irradiâncias e doses de nitrogênio. Agrotrópica. 1998;**10**:21-30

[70] Da Matta FM, Loos RA, Rodrigues R, Barros RS. Actual and potential photosynthetic rates of tropical crop species. Revista Brasileira de Fisiologia Vegetal. 2001;**13**:24-32

[71] Cunningham RK, Arnold PW. The shade and fertilizer requirements of cacao (*Theobroma cacao*) in Ghana. Journal of the Science of Food and Agriculture. 1962;**13**:213-221

[72] Muñoz F, Beer J. Fine root dynamics of shaded cacao plantations in Costa Rica. Agroforestry Systems. **51**:119-130

[73] Bosshart RP, von Uexkull HR. Some occasionally overlooked criteria for assessing fertilizer requirements of high yielding cocoa. In: Seminar on Palm Kernel Utilization and Recent Advances in Cocoa Cultivation; Sawan, Sabah Malaysia. 1987. 29 p

[74] Woodcock TS, Larson BM, Kevan PG, Inouye DW, Lunau K. Flies and flowers. II. Floral attractants and rewards. Journal of Pollination Ecology. 2014;**12**:2014

[75] Sporn SG, Bos MM, Gradstein SR. Is productivity of cacao impeded by epiphytes? An experimental approach. Agriculture, Ecosystems & Environment. 2007;**122**(4):490-493

[76] ten Hoopen GM, Deberdt P, Mbenoun M, Cilas C. Modelling cacao pod growth: Implications for disease control. Annals of Applied Biology. 2012;**160**(3):260-272

[77] Thorold CA. Diseases of Cocoa. Clarendon Press; 1975

[78] Adomako B, Adu-Ampomah Y. Assessment of the yield of individual cacao trees in four field trials. Proceedings of the International Workshop on Cocoa Breeding for Improved Production Systems. 2003;**19**:41-49

[79] Adomako B. Causes and extent of yield losses in cocoa progenies. Tropical Science. 2007;**47**(1):22-25

[80] Entwistle PF. Pests of Cocoa. 1972

[81] Ntiamoah A, George A. Environmental impacts of cocoa production and processing in Ghana: Life cycle assessment approach. Journal of Cleaner Production. 2008:1735-1740

[82] De Souza AL, Delabie JH, Fowler HG. *Wasmannia* spp. (Hym., Formicidae) and insect damages to cocoa in Brazilian farms. Journal of Applied Entomology. 1998;**122**(1-5):339-341

[83] Erickson BJ, Young AM, Strand MA, Erickson EH. Pollination biology of Theobroma and Herrania (Sterculiaceae): II. Analyses of floral oils. International Journal of Tropical Insect Science. 1987;**8**(3):301-310

[84] McKelvie AD. Cherelle wilt of cacao: I. Pod development and its relation to wilt. Journal of Experimental Botany. 1956;**7**(2):252-226

[85] Nichols R. Studies of fruit development of cacao (*Theobroma cacao*)

in relation to Cherelle Wilt: I. Development of the pericarp. Annals of Botany. 1964;**28**(4):619-635

[86] Babin R, Djieto-Lordon C, Cilas C, Dibog L, Mahob R, Bilong CB. True bug (Heteroptera) impact on cocoa fruit mortality and productivity. Journal of Economic Entomology. 2012;**105**(4):1285-1292

[87] Lachenaud P, Mooleedhar V, Couturier C. Les cacaoyers spontanés de Guyane. Nouvelles prospections. Plant Rech. Dév. 1997;**4**:25-30

[88] Uthaiah BC, Sulladmath UV. Cytokinin-like substances and cherelle wilt in cacao (*Theobroma cacao* L). Journal of Plantation Crops. 1980;**8**(2):78-81

[89] Bridgemohan P, Charles RM. A model of the competitive relationships between *R. cochinchinensis* and *Zea mays*. Annals of Applied Biology. 1993;**123**:649-656

[90] Efron Y, Epaina P, Taisa S. Analysis of the factors affecting yield and yield variability in the SG2 cocoa hybrid variety in Papua New Guinea. In: Proceedings of the International Workshop on Cocoa Breeding for Improved Production Systems, Accra, Ghana, 19th-21st October 2003; INGENIC; 2005. pp. 50-61

[91] Lachenaud P. Variations in the number of beans per pod in *Theobroma cacao* L. in the Ivory Coast. II. Pollen germination, fruit setting and ovule development. Journal of Horticultural Science. 1995;**70**(1):1-6

[92] Carr MKV, Lockwood G. The water relations and irrigation requirements of cocoa (*Theobroma cacao* L.): A review. Experimental Agriculture. 2011;**47**(4):653-676

[93] Salef PJM. Growth, flowering and fruiting of cacao under controlled soil moisture conditions. Journal of Horticultural Science. 1970;**45**(2):99-118

[94] Resnik ME, Orchard JE, Mendes LF. The use of Ethrel-induced senescence of cacao fruits to study the control of cherelle wilt. Revista Theobroma. 1980;**10**(4):253-256

[95] Forbes SJ, Northfield TD. Increased pollinator habitat enhances cacao fruit set and predator conservation. Ecological Applications. 2017;**27**(3):887-899

[96] Toledo-Hernández M, Wanger TC, Tscharntke T. Neglected pollinators: Can enhanced pollination services improve cocoa yields? A review. Agriculture, Ecosystems & Environment. 2017

[97] Peleg Z, Blumwald E. Hormone balance and abiotic stress tolerance in crop plants. Current Opinion in Plant Biology. 2001;**14**(3):290-295

[98] Gama-Rodrigues AC, Valle RR, Rossiello ROP. Crescimento, trocas gasosas e relações hídricas de plântulas de cacau em função de diferentes fontes de potássio. Revista Brasileira de Ciência do Solo. 1995;**19**:387-393

Chapter 3

Leaf Cuticular Wax, a Trait for Multiple Stress Resistance in Crop Plants

Kunhikrishnan H. Dhanyalakshmi, Raju Y. Soolanayakanahally, Tawhidur Rahman, Karen K. Tanino and Karaba N. Nataraja

Abstract

Cuticular waxes form the primary interface between a plant and its external environment. The most important function of this hydrophobic interface is regulation of non-stomatal water loss, gas exchange and conferring resistance to a wide range of biotic as well as abiotic stresses. The biosynthesis, transport and deposition of the cuticular waxes are tightly coordinated by complex molecular networks, which are also often regulated in response to various developmental, biotic as well as abiotic cues. Evidences from model as well as non-model systems suggest that targeted manipulation of the molecular regulators of wax biosynthetic pathways could enhance plant resistance to multiple stresses as well as enhance the post-harvest quality of produce. Under the current scenario of varying climatic conditions, where plants often encounter multiple stress conditions, cuticular waxes is an appropriate trait to be considered for crop improvement programs, as any attempt to improve cuticular traits would be advantageous to the crop to enhance its adaptability to diverse adverse conditions. This chapter briefs on the significance of cuticular waxes in plants, its biosynthesis, transport and deposition, its implication on plant resistance to adverse conditions, and the current options in targeted manipulation of wax-traits for breeding new crop types.

Keywords: cuticular waxes, wax biosynthesis, biotic stress, abiotic stress, stress resistance

1. Introduction

In the current era of increasing uncertainties in crop production, emerging constraints and risks demand technical and technological advances in the agricultural sector, and integrative approaches, such as Climate Smart Agriculture (CSA), to address the interlinked challenges of food security and climate change. While maintaining food security is a major challenge for future, the possible solution is to enhance crop productivity along with nutritional security. However, this stance is remarkably limited by the different abiotic as well as biotic environments, where the crops grow and develop.

Drought, excess water (flooding), extremes of temperatures (cold, chilling, frost, and heat), salinity, high and/or low light, mineral deficiency, and toxicity are the common abiotic stresses for crop production. These stresses alter plant metabolism, growth, development, and in extreme cases cause the cessation of vegetative and reproductive growth. Some of the abiotic stresses such as drought, high temperature and salinity can influence the occurrence and spread of biotic agents like pathogens, insects, and weeds [1]. In crops like tomato, cucurbits and rice, temperature is one of the most important deciding factors for the occurrence of bacterial diseases [2]. Temperature can also alter the incidence of vector-borne diseases by modifying spread of vectors.

But, in their natural environment, plants face combination of stresses, especially under the changing climate scenario. The effect of stresses would be more pronounced under combined (biotic and abiotic) stresses [3], while simultaneous occurrence of abiotic and biotic stresses are more destructive to crop production [4]. Hence, there exists a need now, to look for common traits that can contribute for plant adaptation to such multifarious stressful conditions and sustain crop productivity as well. In this scenario, it is desirable to have a single trait that can confer tolerance to multiple (abiotic and biotic) stresses. Cuticular waxes, a major component of plant cuticle covering all the aerial parts of the plants, can be considered as an important trait for combined stress resistance.

2. Cuticular waxes: a component of plant cuticle

The cuticle is a unique structure developed by land plants during the course of their evolution from an aquatic to a terrestrial lifestyle [5]. The primary role of this lipophilic layer, comprising cutin and cuticular waxes, was to limit non-stomatal water loss by functioning as a physical barrier between the plant surface and its external environment [6]. Development of a cuticular barrier is one of the major adaptive mechanisms for survival and growth of plants under water limiting terrestrial conditions [7]. As the primary barrier between the aerial surface of plants and the external environment, the cuticle also protect the plants from mechanical rupture or injury, toxic gases and ultra violet radiation [8–10]. The cuticle also has notable roles associated with growth and developmental processes like preventing epidermal fusion by establishing normal organ boundaries [11], and phytohormone homeostasis [12]. The cuticle and its components are known to play essential roles as signaling molecules for pathogens and for the plants themselves [13]. Another important role is in fruits, where it influences quality, defense and post-harvest shelf life [14]. In fruits, water retention [15] firmness [16] and its responses to physical and biotic stresses are also influenced by the cuticle [17].

The cuticle is composed of a covalently linked scaffold of cutin and a mixture of soluble cuticular lipids (SCL), called as waxes [10, 18]. Structurally, cutin is made of covalently cross linked C16 or C18 oxygenated fatty acids and glycerol, forming the most abundant structural component of the cuticle [19]. The waxes within the cuticle function as an actual barrier against the diffusion of water or solutes [20, 21]. The waxes occur in two layers; forming two distinct physical layers called intra- and epi-cuticular waxes [22]. The former is dispersed within the cutin polymer while the epi-cuticular wax is deposited on the outer surface as crystals or films [22, 23]. This outermost layer can be physically stripped off the surfaces using aqueous glue [23, 24]. These waxes are composed of a variety of organic solvent-soluble lipids; consisting of very-long-chain fatty acids (VLCFA) and their derivatives. The major composition of VLFCAs are alkanes, wax esters, branched alkanes, primary alcohols, alkenes, secondary alcohols, aldehydes ketones, and unsaturated fatty alcohols, as well as cyclic compounds including terpenoids and metabolites such as sterols and flavonoids [19, 25–27]. Wax composition varies with crop species and differs in their functions and responses to biotic and abiotic environments [10].

As per recent studies, intra-cuticular waxes form the primary transpirational barrier and the contribution of epi-cuticular waxes as a transpirational barrier depends on the species-specific cuticle composition [28]. In species like *Tetrastigma voinierianum*, *Oreopanax guatemalensis*, *Monstera deliciosa*, and *Schefflera elegantissima*, intra-cuticular wax pre-dominantly act as a transpirational barrier while in *Citrus aurantium*, *Euonymus japonica*, *Clusia flava*, and *Garcinia spicata*, both intra- as well as epi-cuticular waxes had equal contribution as transpirational barriers [28]. A study from *Prunus* suggests that intra-cuticular waxes of the cuticle form the actual transpirational barrier [29] and not epi-cuticular waxes [30].

3. Ecological significance of cuticular waxes

The cuticular waxes confer diverse surface properties to plant parts, which actually play the key role in controlling non-stomatal water loss and gas exchange, and protection from external environment. Leaf cuticular wax amount and crystal morphology regulated post-harvest water loss from leaves [31]. Epi-cuticular wax films give glossy appearance to leaves and fruits, while wax crystals (β-diketones) conferred dull, glaucous appearance to leaves and stems [10]. The thickness [5] composition and properties of the waxes vary with crop species and are found to be induced under diverse stressful conditions [32]. These differences reflect their functions and responses to biotic and abiotic environments [10]. Importance of cuticular wax accumulation in plant resistance to both biotic as well as abiotic stress conditions is now well documented [12, 33, 34].

3.1 Abiotic stresses

One of the most important roles of the waxes is to protect the plant surfaces from excessive solar and ultraviolet (UV) radiations. Cuticular waxes scatter UV-B radiation [35] and was demonstrated in apple [36]. As per studies from model systems as well as crops, increased cuticular wax biosynthesis improves drought stress resistance [37]. In rice, wheat, barley and sorghum, grain yield under water limiting conditions have positive correlation with wax content [38–41] . Hence, in crop plants, higher cuticular wax content is a promising trait for stress resistance as well as yield under water limiting conditions [27]. In mulberry, increasing wax load is useful to manage post-harvest water losses [42]. In barley, cuticular wax components act as a barrier to water loss and contribute to salt stress resistance [43]. Heat stress resistance is also positively correlated with wax accumulation in bahia grass [44]. Under heat stress, the wax load in sorghum was correlated with its ability to maintain the canopy temperature cool, resulting in reduced water loss [45]. Similarly, pea varieties with thicker wax load also exhibited lower canopy temperature, thereby limiting water loss and associated heat stress [46].

Cuticular waxes play an important role in preventing non-stomatal water loss during drought and high temperature stress, as well as enabling frost avoidance. Such climatic stressors can induce a heavier wax load and change the chemical composition of waxes by accumulating longer aliphatic compounds on plant tissues [47]. Drought increases stiffness and quality of the plant cuticle under climate change [48]. Similarly, the leaf cuticular surface is the first barrier blocking destructive ice penetration into the leaf cells in freezing avoidance mechanisms [49]. Using a hydrophobic film, Wisniewski et al. [50] showed the importance of the epi-cuticular hydrophobicity enabling avoidance of freezing in sensitive plants. The critical nature of the cuticular layer in frost avoidance of corn is also clearly demonstrated [51]. Freezing avoidance is the only mechanism of frost resistance in sensitive plants. In fact, the first demonstration of a transgenic organism in agriculture was the alteration of the cell wall protein secondary structure on ice nucleating bacteria, *Pseudomonas syringae* and *Erwinia herbicola*, which then prevented ice nucleation across the cuticle and avoided leaf damage [52, 53]. In future, injury due to frost stress will be more, not less under global warming [54]. Hence, a better understanding of stress-induced wax modification among crop plants holds promise to cope with climate change.

3.2 Biotic stresses

The cuticle and its components act as signaling molecules to favor fungal growth and development, and infections in plants [55, 56]. Surface waxes act as cues to activate fungal developmental processes like appressorium formation, pre-penetration processes, etc., in crop plants like avocado, wheat, rice, maize and peanut [13, 57–59]. However, the hydrophobic nature of the cuticle also renders it a barrier for bacterial as well as fungal pathogens [60], a desirable trait for disease resistance. Waxes are known to protect lotus from pathogen infection [61]. It repulses pathogen spores and atmospheric pollutants like acid rain and ozone [32]. Another role of waxes is in plant-insect interaction; to attract or to serve as a deterrent [62]. It prevents insect attachment to plant surface oviposition and feeding [63, 64] and hence confer tolerance to insects in crop plants [65, 66].

4. Molecular biology of cuticular wax biosynthesis and deposition

Studies in *Arabidopsis* and subsequently, barley, rice and tomato systems have significantly contributed for the elucidation of the complex regulatory pathways underlying the biosynthesis, transport and deposition of wax components on plant surfaces [26, 27, 67]. Cuticular wax biosynthesis predominantly occurs in epidermal cells. The biosynthetic pathway initiates exclusively in the outer membranes of the plastids of epidermal cells where C16 and C18 fatty acids are synthesized, exported to the cytosol as acyl-CoAs and then elongated up to C34 at the endoplasmic reticulum (ER); through a series of enzymatic reactions [19, 26]. The synthesized components are subsequently transported through the apoplastic pathway and deposited on the cuticle. The key steps involved [32] are summarized here.

4.1 Synthesis of malonyl-CoA

The *de novo* fatty acid biosynthesis initiates with the synthesis of malonyl-CoA. It is initiated with the transfer of a bicarbonate derived CO_2 molecule to the biotin moiety of a biotin carboxylate carrier protein (BCCP), that form N-1,2 carboxybiotin biotin carboxylate carrier protein-BCCP. The reaction is catalyzed by biotin carboxylase (BC). The CO_2 is further transferred to acetyl-CoA by carboxyltransferases (CT). Acetyl-CoA carboxylase (ACCase), a multifunctional enzyme system then catalyzes the formation of malonyl-CoA, from acetyl-CoA [32], which will be subsequently used for *de novo* fatty acid biosynthesis.

4.2 De novo fatty acid biosynthesis

De novo synthesis of acyl chain in the stroma of plastids is catalyzed by a series of enzymatic steps, which collectively forms fatty acid synthase complex (FAS). The series of reactions with the catalyzing enzymes are:

a. Condensation of malonyl-acyl carrier protein (manolyl-ACP) with acetyl-CoA to form 3-ketoacyl-ACP catalyzed by β-ketoacyl-ACP synthase (KAS III).

b. Reduction of 3-β-ketoacyl-ACP to 3-hydroxyacyl-ACP, catalyzed by 3-βketoacyl-ACP reductase.

c. Dehydration of 3-hydroxyacyl-ACP to *trans*-Δ^2-enoyl-ACP, catalyzed by β-hydroxy acyl ACP dehydratase.

d. Reduction of *trans*-Δ^2-enoyl ACP to Acyl-ACP by Enoyl ACP reductase.

This complex also includes an acyl carrier protein (ACP), a cofactor component of FAS to which the growing acyl chain remains esterified. These sequential reactions result in a fully reduced acyl chain, extended by two carbons in each cycle [68] through the sequential round of condensation, reduction, dehydration and second-reduction steps [69]. Repetition of the cycle for six times generates palmitoyl-ACP (16:0-ACP), where the condensation reactions are catalyzed by KAS I. One final cycle reaction between palmitoyl-ACP and malonyl-ACP utilizes KAS II to generate stearoyl-ACP (18:0-ACP). These products are further processed by stearoyl-ACP desaturase (introduce double bonds), plastidial acyltransferases, and acyl-ACP thioesterases (hydrolases). The fatty acyl-ACP thioesterases (FATA and FATB) hydrolyzes the C16-C18 acyl-acyl carrier proteins to generate fatty acids, which are then exported out of the plastids to undergo modifications in the ER [69].

4.3 Elongation of fatty acids

The C16 and C18 compounds, hydrolyzed by acyl-ACP thiosterases are activated into C16- and C18-CoA by long chain acyl-CoA synthetases (LACSs) and exported to the ER. The C16 and C18 acyl-CoA then act as a substrate for fatty acid elongase (FAE) complex, localized on the ER, which adds two carbons successively to form VLCFAs with C26-C34 chains. FAE complex are heterotetramers of independently transcribed, monofunctional proteins. They operate a reiterative cycle of four reactions catalyzed by

i. β-Ketoacyl-CoA synthase (KCS) that catalyze the two carbon condensation to acyl-CoA.

ii. β-Ketoacyl-CoA reductase (KCR) that catalyze the reduction of β-ketoacyl-CoA.

iii. β-Hydroxyacyl-CoA dehydratase (HCD) that catalyze the dehydration of β-hydroxyacyl-CoA.

iv. Enoyl-CoA reductase (ECR) that reduces the enoyl-CoA ultimately leading to VLCFAs [69–71].

4.4 Wax biosynthetic pathways

The elongated products are further modified to produce wax components i.e., to primary alcohols, alkyl esters, aldehydes, alkanes, secondary alcohols, ketones and free fatty acids, via two pathways (i) acyl reduction pathway (generates primary alcohols and wax esters) and (ii) decarbonylation pathway (generates alkanes, aldehydes, secondary alcohols, and ketones).

i. Acyl-reduction pathway: fatty acyl-CoAs are converted into primary alcohols catalyzed by fatty acyl-CoA reductase (FAR) through an intermediate aldehyde [71]. A bi-functional wax synthase/acyl-CoA:diacylglycerol acyltransferase (WS/DGAT) enzyme, WSD1 condenses the generated fatty alcohols and C16:0 acyl-CoA into wax esters [26].

ii. Decarbonylation pathway: acyl-CoAs are reduced to aldehyde intermediate by FAR, which are subsequently decarbonylated into alkanes, catalyzed by aldehyde decarbonylase. Stereospecific hydroxylation of alkanes catalyzed by midchain alkane hydroxylase 1 (MDH1) give rise to secondary alcohols, and oxidation of these alcohols form corresponding ketone [32]. Additional hydroxylation and oxidation reactions lead to the esterification of secondary alcohols with fatty acids and formation of diols, hydroxyl ketones and diketones [32].

4.5 Transport and deposition of cuticular waxes

The wax components generated are then transferred from the ER to the plasma membrane (PM) through Golgi and trans-golgi network mediated vesicle trafficking or non-vesicular trafficking [72]. Further, adenosine triphosphate binding cassette (ABC) transporters in the plasma membrane (homodimers and heterodimers) export the wax components to the epidermal surface [73]. Lipid transfer proteins (LTPs) like glycosylphosphatidylinositol (GPI)-anchored LTPs (LTPGs), attached to the outer surface of the plasma membrane are also

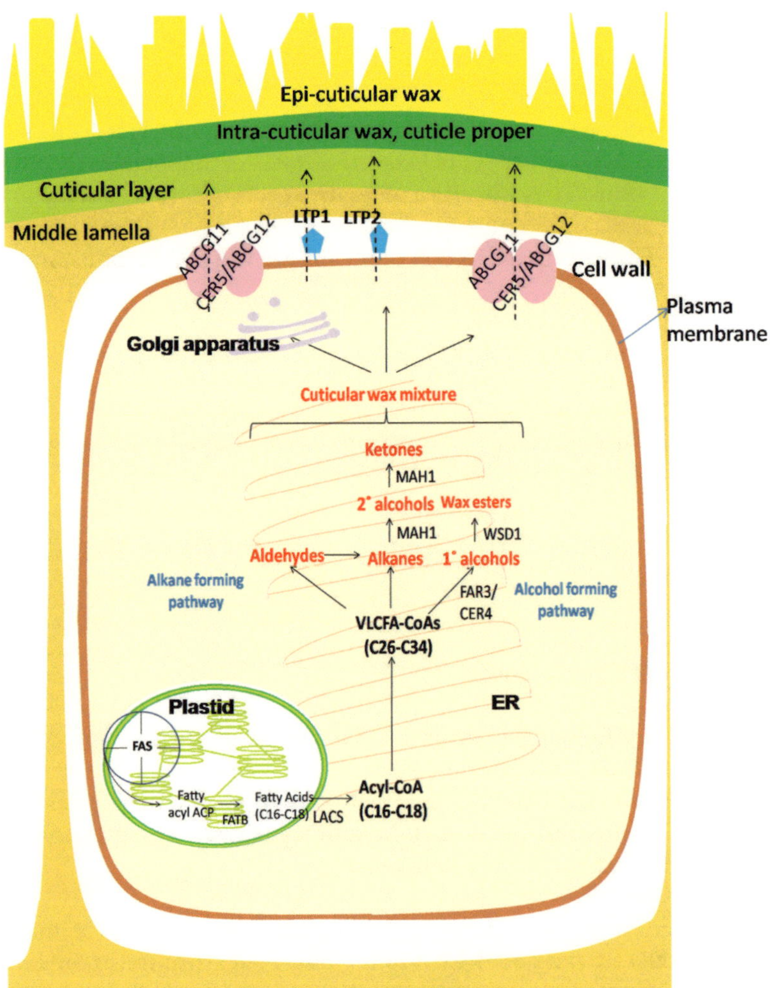

Figure 1.
Schematic representation of wax biosynthesis, transport and deposition in plants.

Gene	Protein type	Role	Reference
Cuticular wax biosynthesis			
ACC1	Acetyl CoA carboxylase	Synthesis of malonyl CoA substrates	[79]
FATB	Acyl acyl carrier protein thioesterase	Supply of saturated fatty acids for wax biosynthesis	[80]
CUT1 /CER6/KCS6	VLCFA condensing enzyme (β-ketoacyl-CoA synthase)	Regulation of VLCFA biosynthesis/elongation of 24C fatty acids	[81]
CER1/CER22	Aldehyde decarbonylase	VLC alkane biosynthesis	[82]
KCS1	β-ketoacyl-CoA synthase	Elongation of 24C fatty acids	[83]
KCS20; KCS2/DAISY	3-ketoacyl-coenzyme A synthase	Required for VLCFA elongation to C22	[84]
LACS1/CER8; LCAS2	Long chain acyl CoA synthetase	Synthetase activity for VLCFAs C20-C30	[85]
KCS9	3-ketoacyl-coenzyme A synthase	Elongation of C22-C24 fatty acids	[86]
WAX2/YRE/FLP1/CER3	Aldehyde-generating acyl-CoA enzyme	Required for synthesis of aldehydes, alkanes, secondary alcohols, and ketones; biosynthesis of cuticular membrane	[76, 87]
CER10	Enoyl-CoA reductase	Biosynthesis of VLCFA	[88]
CER4/FAR3	Alcohol forming fatty acyl CoA reductase	Formation of C24:0 and C26:0 primary alcohols	[89]
CYP96A15 (cytochrome P450 enzyme)	Midchain alkane hydrolase	Formation of secondary alcohols and ketones (stem cuticular wax)	[78]
WSD1	Wax ester synthase/diacylglycerol acyltransferase	Wax ester biosynthesis	[90]
PASTICCINO2 (PAS2)	3-hydroxy-acyl-CoA dehydratase	VLCFA synthesis in association with CER10, an enoyl-CoA reductase	[91]
KCR1	β-Ketoacyl-CoA reductase	Required for VLCFA elongation	[70]
CER2	BAHD acyltransferase	Fatty acid elongation beyond C28	[92]
CER17 (ECERIFERUM1)	Acyl-CoA desaturase like 4	n-6 desaturation of very long chain acyl-CoAs	[93]
Transport and deposition			
AtWBC12/CER5	ATP binding cassette (ABC) transporter	Transport of cuticular waxes	[94]
LTPG1	Lipid transport protein	Cuticular wax export or accumulation	[74]
ABCG11/WBC11/DESPERADO	ATP binding cassette (ABC) transporter	Secretion of surface waxes in interaction with CER5	[73, 95]
LTPG2	Lipid transport protein	Cuticular wax export or accumulation	[96]
GLN1, ECH		Vesicle trafficking	[72]

Table 1.
Key genes involved in wax biosynthesis, transport and deposition identified from the model system Arabidopsis.

directly or indirectly involved in wax export [74]. A brief representation of wax biosynthesis, transport and deposition with key genes, is presented in **Figure 1** (adapted from [19, 26, 27, 32, 69, 71]).

Early studies in barley mutants with little or no wax on aerial plant parts, called glossy or glaucous were termed as eceriferum (cer), where cera means wax and ferre means to bear [75]. Subsequently, the wax defective mutants in *Arabidopsis* with bright, shiny, or glossy stems or leaves were also termed as *eceriferum* (*cer*) [76]. The wax locus from maize and *Brassica napus* is termed as *glossy* [68]. With the help of forward genetic screens using wax defective mutants and reverse genetic approaches [77, 78], considerable progress has been achieved in understanding wax biosynthesis, transport and deposition. **Table 1** gives an overlook of the key genes involved in wax biosynthesis, transport and deposition identified from the model system *Arabidopsis.*

5. Regulation of cuticular wax biosynthesis

While the complex wax biosynthesis and transport pathways are well determined, the information on underlying regulatory mechanisms is still fragmentary. There is limited information that these processes and their candidate pathway genes are influenced by developmental factors. The cuticle development is an intrinsic part of cell developmental processes like organ development, cell partitioning, etc. [11]. PAS2, acy-CoA dehydratase, regulating the synthesis of VLCFA during wax biosynthesis in the epidermis is essential for proper cell proliferation during development [97]. Wax deposition is also known to occur in an organ-specific manner during its development and is influenced by diverse environmental conditions as well [17]. The available information on the exact developmental regulation of wax biosynthesis is however, limited. As per evidences from leek (*Allium porrum* L.), wax accumulation and elongation activities are highly induced within a defined and an identifiable region of leaf [98]. The expression of plastidial fatty acid synthase (FAS), FAEs that regulate elongation of long-chain fatty acids in the microsomal membranes and acyl ACP-thioesterases are probable targets of developmental regulation, depending upon the need to produce fatty acid precursor pools [98]. Some of the key genes involved in wax biosynthesis are also affected by defects in the organization of organelles, especially the ER. A mutation of PEX10 (peroxisome biogenesis factor 10) in *Arabidopsis*, which disrupted the ER network, in turn lead to mislocalization of CER4, CER1, SHN1 and WAX2, affecting cuticular wax biosynthesis [99].

There is increasing evidence to show that wax biosynthesis and its pathway genes are regulated at transcriptional, post-transcriptional and translational levels [26, 100]. A wide range of abiotic factors like light, water, temperature, salinity etc., influence wax biosynthesis and deposition. An increase in cuticular wax content is observed in bean, barley and cucumber on exposure to UV-B light [101]. In cotton, enhanced UV-B radiation specifically increased the epicuticular wax load on the adaxial surface of leaves [102]. There is an also an up-regulation of wax biosynthetic genes in salt tolerant rice genotypes under stress [103]. Although the underlying mechanisms have not been well explored in the above conditions, there is sufficient information on the influence of drought or moisture stress on wax biosynthesis in plants. A significant increase in wax load in *Arabidopsis* plants subjected to water stress is indicative of its regulation under drought [17]. In crops like rice, wheat, tobacco, alfalfa, peanut and cotton, etc., an increase in cuticular wax accumulation was observed under moisture stress condition [104]. Drought induced accumulation of wax biosynthesis is positively correlated with drought tolerance in crops like oat, rice, wheat and forage crops, etc. [104–107].

The transcript levels of several genes involved in wax biosynthetic pathways are regulated in response to abiotic stresses. FAR5, a fatty acyl CoA reductase, in wheat responsible for accumulation of long chain primary alcohols of C26:0, C28:0 and C30:0 are regulated by drought, ABA and cold [108]. The transcripts of KCS2/DAISY, a 3-ketoacyl-coenzyme A synthase required for the elongation of VLCFA are up regulated under water deficit conditions [84]. Osmotic stress induces the expression of CER1, that regulates alkane biosynthesis; while the over expression of CER1 increased susceptibility to bacterial and fungal pathogens [109]. Hypoxia is also known to affect total wax loads on *Arabidopsis*. The expression of KCS, KCR1, ECR/CER10 and PAS2, components of fatty acid elongase complex in *Arabidopsis* stem and leaves is affected which in turn affects the production of VLCFA precursors of wax biosynthesis. The wax synthesis genes like MAH1, CER3, CER4, WSD1, etc., and several genes associated with wax and lipid transport are also affected by hypoxia [110]. There is also indication on the regulation of wax biosynthesis in response to cold. Acteyl-CoA carboxylase plays the essential role for cold acclimation in *Arabidopsis*. In sensitive to freezing3 (*sfr3*) mutants, with a missense mutation in ACC1, the long chain components of leaf cuticular wax were reduced and there was inhibition on the wax deposition on inflorescence stem, which rendered the plants sensitive to cold stress [111]. Wax biosynthesis is also reported to be regulated in response to carbon dioxide (CO_2) concentration. This is mediated by HIC (High Carbon Dioxide), a gene encoding a 3-keto acyl coenzyme A synthase (KCS)-an enzyme involved in the synthesis of very-long-chain fatty acids that influences stomatal development in *Arabidopsis* [112].

With the identification of several transcription factors (TFs), transcriptional regulatory mechanisms are considered to be a major contributor for the wax biosynthesis [113]. WIN1/SHN1 (WAX INDUCER 1/SHINE1) is a TF from AP2/EREBP family initially reported to regulate cuticular wax and then cutin biosynthesis by regulating the expression of *CER1*, *KCS1*, *CER2*, *LACS2*, *GPAT4*, *CYP86A4*, *CYP86A7* and *HTH*-like genes [114]. SHN1 overexpression increased drought tolerance in *Arabidopsis* [115]. Wax synthesis regulatory gene 1 (WR1) from rice [116] and SHN1 from wheat [117], both homologs of WIN1/SHN1 from *Arabidopsis* also reduced water loss and improved drought tolerance. Transcriptional repression by diurnally controlled DEWAX2 is another important for regulator of wax biosynthesis in *Arabidopsis*. Compared to wild type, the total wax loads in *dewax2*, were increased by 12 and 16% respectively in rosette and cauline leaves [118, 119]. Another candidate from AP2/ERF TF family, WRINKLED4 (WRI4) positively regulates wax biosynthesis in stems. *wri4* mutants expressed 28% reduction of total wax loads in stems, although siliques and leaves were unaffected. Hence WRI4 act as a transcriptional activator to regulate the expression of LACS1, KCR1, PAS2, ECR and WSD1, to maintain the levels of 29C long alkanes, ketones and secondary alcohols in stems [113]. MYB94, regulate the expression of wax biosynthetic genes like WSD1, KCS2/DAISY, CER2, FAR3 and ECR to activate cuticular wax biosynthesis and is up regulated by drought and ABA. This also conferred tolerance to drought stress in *Arabidopsis* and *Camelina* [120]. MYB96, an ABA responsive TF also regulates wax biosynthesis under drought [121]. In *Camelina*, MYB96 activated the expression of wax biosynthetic genes KCS2, KCS6, KCR1-1, KCR1-2, ECR, and MAH1 which resulted in high levels of alkanes and primary alcohols and improved drought tolerance [120]. MYB96 acts as a component of plant disease resistance, through salicylic acid mediated signaling [122]. Both MYB94 and MYB96 share a common region containing MYB consensus motifs in the promoter of their target wax biosynthetic genes [123]. Hence MYB94 and MYB96 have an additive role on plant cuticular wax biosynthesis and under drought and ABA conditions.

In addition, to transcriptional regulation, wax biosynthesis is regulated by other events. Expression of CER3/WAX2/YRE, an aldehyde-generating acyl-CoA enzyme

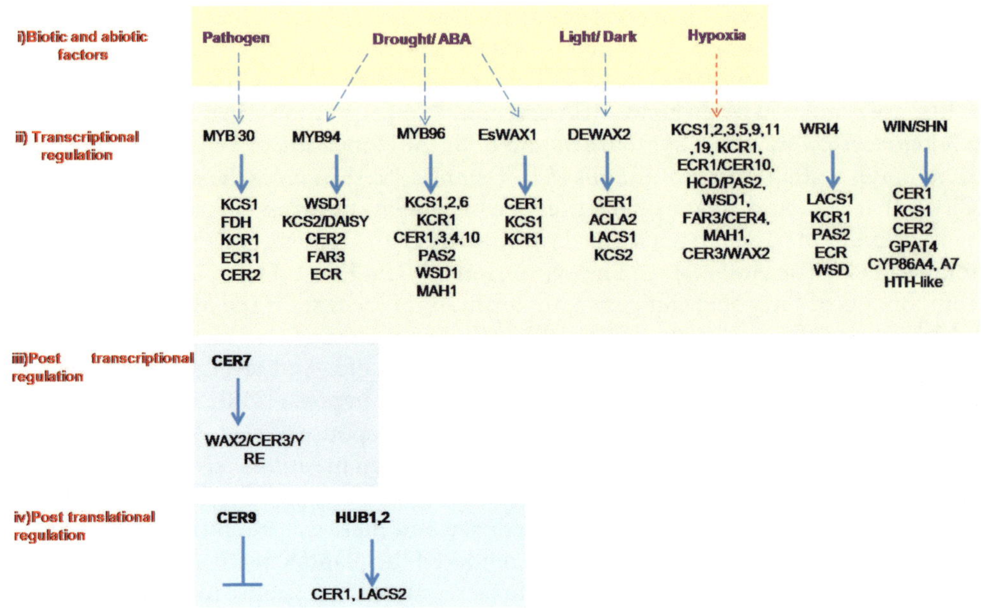

Figure 2.
Brief representation of the key regulatory events in wax biosynthesis and their targets.

in the wax biosynthetic pathway is regulated by CER7, a core RNA processing and degrading exosomal subunit. CER7 regulates WAX2 transcript levels by degrading a specific mRNA species encoding its negative regulator [124]. Many of such regulators have been identified from model systems as well as crop species and a brief overview of the key regulatory events and their targets has been presented in **Figure 2**.

6. Cuticular wax trait in imparting stress resistance

Under field conditions, crops encounter multiple biotic and/or abiotic stresses simultaneously at different stages of developments. Cuticular waxes have a direct role in multiple stress tolerance in crops [109]. In cucumber, wax biosynthesis has been shown to have key roles in influencing the plant responses to biotic as well as abiotic stresses [125]. In sorghum, genes regulating leaf waxes have critical role in regulating tolerance to drought and heat stress [45]. Considering the relevance of cuticular waxes under diverse biotic as well as abiotic stressful conditions, as discussed above and under combined stress conditions, it can be an ideal trait to tackle multiple stresses in crop plants.

6.1 Biotic stresses

6.1.1 Pathogens

Being the outermost layer of plant cuticle, the epi-cuticular wax can serve as a first line of physical defense against pathogens and herbivores. However, increasing thickness and hydrophobicity of the cuticle through over-deposition of the wax may not necessarily increase the resistance of the plant against biotic stresses. The composition and structure of wax in the cuticle can constitute the source of signals for the foreign invaders and for the plants themselves. Thus, the roles of cuticular wax could be multifunctional and can vary not only for various plant species but also for different kinds of pathogens. Functional study of the *DEWAX* gene, a negative regulator of wax biosynthesis in *Arabidopsis*, is a good example of this complexity. The *dewax* mutant line in *Arabidopsis*, with increased epicuticular wax and decreased cuticular permeability, showed susceptibility to the fungal pathogen *Botrytis cinerea*, but resistance to the bacterial pathogen *Pseudomonas syringae* [126]. Moreover, *DEWAX* overexpressing lines in *Arabidopsis* and *Camelina* showed inverse defense modulations to *B. cinerea* and *P. syringae* as compared to *dewax* mutant in *Arabidopsis* [126].

Wax and cutin components in the plant cuticle could function in pattern- and effector-triggered immunity (PTI and ETI) and could serve to generate local and systemic acquired resistance against numerous pathogens [127]. During plant-pathogen interaction, the plant cuticle can be affected by enzymes synthesized and secreted by the pathogens. Many fungal pathogens synthesize and secrete hydrolytic enzymes (for example, cutinases, esterases and lipases) at the early stage of infection that directly target the cuticle [128–131]. *Fusarium oxysporum* secretes cutinases that degrade cutin layers in the cuticle and generates cutin monomers that support fungal adherence to the host plant and facilitate the initiation of infection [128]. Hexadecanediol, a cutin component in rice can facilitate spore germination and differentiation for pathogenic fungi *Magnaporthe grisea* and *B. cinerea* [55]. Presence of a very-long-chain C_{26} aldehyde (a wax component) was important for the barley powdery mildew fungus (*Blumeria graminis*) to initiate infection in host plant species. Germination and appressorial differentiation of *B. graminis* were strongly prohibited in aldehyde free *glossy11* mutant in corn. Spraying of *n*-hexacosanal (C_{26}-aldehyde) or wax preparation from wild-type corn can restore the conidial formation and differentiation [59].

Plant can also recognize the attachment of pathogens and activate defense responses against them, in which pathogen-infection generated plant products, such as cutin monomers or cell wall oligosaccharides, can act as signaling molecules [132]. Defense responses in plants are often manifested as alternations of the cuticle. *Colletotrichum acutatum* infection in citrus resulted in increased lipid synthesis in the epidermal cell and increased deposition of those lipids in cuticle, the process eventually changes the structure of the cuticle [133]. Cuticular biosynthesis was also found to be up-regulated in tomato fruit following infection by fungal pathogen *C. gloeosporioides* [134].

Cuticular permeability plays a vital role in almost all plant-pathogen interactions. A more permeable cuticle can lead to either resistance or susceptibility to pathogens. Elevated deposition of cuticular wax as well as the presence of hydrophobic wax components (e.g., very-long-chain alkanes or ketones) can make a cuticle less permeable. Mutation or overexpression of genes that diminish biosynthesis of various wax components can generate the opposite effect. There are number of wax-deficient mutant and transgenic lines in *Arabidopsis* and other plant species with diminished cuticular permeability showed resistance to the fungal pathogen *B. cinerea* [34, 127]. However, the phenomenon is not true for all wax deficient plant lines. Wax and cutin deficient *acp4* and *gl1* mutants in *Arabidopsis* displayed increased sensitivity to *B. cinerea* [135, 136]. Mutations in SHINE transcription factors in other studies also showed alteration in cuticular wax accumulation, and susceptibility to *B. cinerea* infection [137, 138].

6.1.2 Insects and herbivores

Epicuticular wax also plays important roles in plant interaction with insects and herbivores. Flowering plants have evolved with cuticular wax of various forms, sizes and structures that are either enabling the attachment and movement of pollinating insects, or reducing the attachment of herbivorous insects and pests on the plant surfaces. Reducing the attachments of herbivores on plant surfaces is a part of a plant defense strategy against herbivores.

Most plant body surfaces are covered with a two-dimensional (2D) epicuticular wax film of various thicknesses. In many species, wax film is protrudes with three-dimensional (3D) wax crystals. Wax crystals can generate various shapes as revealed by electron microscopic analysis, such as rodlets, threads, platelets and tubules [61]. The complexity of these various shapes originates from the molecular self-assembly of various wax components, in which morphology of those crystals is also correlated with the presence of specific chemical components in the wax [139, 140]. Many experimental studies and reports from various plants species (for example, from genera *Eucalyptus*, *Pisum*, *Brassica*) have shown that 3D wax crystals have protective functions against insects, in general, including the herbivorous insects [141]. Studies with *Eucalyptus* species in canopy found that glaucous juvenile leaves containing high quantities of wax crystals were less prone to herbivorous infestation as compared to the glossy adult leaves [142]. Feeding rates of flea beetles, *Phyllotreta cruciferae*, on low-wax glossy (*eceriferum, cer*) *Brassica napus* mutant lines were much higher as compared to the wild-type *B. napus* [143]. Cuticular surfaces with wax crystals also interferes with the attachment, locomotion and foraging behavior of predatory insects and parasitoids [65, 144]. *Pisum sativum* lines with higher prevalence of crystalline epicuticular wax (CEW) were found more favorable for four predatory coccinellid species to attach, move and consume more aphids as compared to the *P. sativum* mutant line with reduced CEW [145]. Flowering stems with high CEW of numerous other plant species (for example, species under the genera *Salix*, *Hypenia*, *Eriope*) often generate slippery surfaces that prevent the movement of nectar robbers, ants and other plant pests [141, 146].

Several hypotheses have been proposed and tested on the mechanisms of wax crystal inhibition of insect attachment inhibition: (i) roughness hypothesis; (ii) contamination hypothesis; (iii) fluid absorption hypothesis [141]. Wax crystals, in general, generate a micro-rough surface on the cuticle that may prevent adhesive pads of the insects to stick, preventing them to successfully attach to the plant surface [144, 147, 148]. Contamination hypothesis proposed that detached wax crystals of the cuticular surface of some plants can adhere to the insect attachment organs (e.g., adhesive pads), contaminate those, and as such subsequent insect attachment becomes challenging and unsuccessful [147–149]. Adhesive pads of many insects secret fluids, which can also enhance wax crystal contamination to attachment organs. Fluid secretion from the adhesive pads are supposed to help insects to pursue successful attachment to the plants. However, there is evidence certain plant species have crystalline wax coverage that can absorb the fluids secreted by the adhesive pads and prevent the insects to successfully attach to the cuticle [150, 151].

The study of cuticular wax involvement in biotic stress resistance is complex with a multitude of organisms spanning insects to disease. The story is still not clear and field situations in which interactions between organisms and abiotic stresses and the role of cuticular wax needs to be evaluated. Nevertheless, certain consistencies are evident in that permeability of the cuticular layer appears to be important in pathogen invasion and wax crystals play an important role in insect intervention by the cuticular layer. These areas of research merit further investigation.

6.2 Abiotic stresses

As mentioned above, abiotic stresses such as drought, extremes of temperatures, salinity, etc., cause significant losses in crop productivity. Since most of the stresses occur simultaneously, crop breeders are looking for traits contributing for multiple stress resistance. From this context, cuticular wax can serve as ideal trait. Drought stress, a major abiotic stresses in tropical regions, influences the biosynthesis and composition of cuticular wax in crops [27]. The importance of cuticular wax in desiccation tolerance is evident that, compared to gymnosperms and angiosperms, many early extant plants such as ferns, and horsetails are more sensitive to dehydration [152]. In crops like pea, cuticular wax load increases when subjected to drought stress [46]. In rice, *gl1*-1/wsl2 and *gl1-2* loss-of-function mutants with reduced wax load exhibited sensitivity to drought compared to the wild type plants [104, 153]. Drought stress is known to increase the wax content and alter composition of cuticular wax in many plants such as pea [46], *Arabidopsis* [17, 115], tobacco [154], alfalfa [155]. Significant correlations between the wax content and yield, drought tolerance and water-use efficiency have been reported in different crops such as sorghum [38], barley [156], rice [41], and wheat [157, 158]. These reports demonstrate that less wax or non-waxy crops/genotypes are sensitive to desiccation with poor drought-tolerance compared to the crops having more cuticular wax [105]. The existing evidences suggests cuticular wax is responsible for reducing non-stomatal transpiration by increasing cuticular resistance [43]. The cuticular waxes also have roles in imparting resistance to salinity stress, mainly by regulating residual transpiration. A significant negative correlation observed between residual transpiration and total wax content, reports residual transpiration could be a fundamental mechanism by which plants optimize water-use efficiency under salinity stress [43]. As discussed above, wax accumulation also correlated with high temperature resistance in plants [44]. Leaf surface waxes help to maintain cooler canopy in sorghum under heat stress [45]. The cuticular waxes can further help in protecting plants from high light stress [101]. The cuticular wax has a role in protecting plants from excessive ultraviolet (UV) light and there are reports indicating that elevated UV-B radiation can affect plant cuticular wax formation [101, 159, 160]. Based on the existing information, as mentioned above, cuticular wax, can be treated as the first protective layer and an important trait contributing for both biotic and abiotic stresses.

7. Attempts by crop biologists to manipulate cuticle traits

7.1 Breeding

Identification of genomic regions contributing wax traits is crucial in manipulating wax characteristics using breeding approaches. In rice, quantitative trait loci (QTL) linked to the leaf epi-cuticular layer was identified corresponding to EM15_10-ME8_4-R1394A-G2132 region on chromosome 8 [161]. In sorghum, a crop with the ability to produce profuse amounts of EW, <u>BL</u>OO<u>M</u>-<u>C</u>UTICLE (BLMC) locus from chromosome 10, was identified to account for profuse wax production. BLMC region corresponds to approximately 153,000 bp with three co-segregating markers and an acyl CoA oxidase with seven other putative candidates. BLMC mutation affected C28-C30 free fatty acid fractions and hence cuticle properties in culm and leaves, disrupted EW production and increased plant death rating in field at anthesis [162]. With the genetic analysis of F2 population from HUAYOU2 (P1 X M36), BoWax1 locus (*Brassica oleraceae* Wax 1) is identified to be controlling glossy green trait in cabbage, due to a deletion mutation of two nucleotides in the cDNA of Bol013612 of HUAYOU2. BoWax1 locus maps to chromosome C01 [163]. The wax biosynthetic pathway genes identified in pearl millet were co-located to the QTL controlling biomass production under early drought stress and stay green traits [164]. Targeted breeding using the modern molecular breeding for this trait would be useful.

7.2 Transgenic

With the elucidation of wax biosynthetic pathways and identification of key regulators, attempts were made in crop plants to engineer cuticle properties and to enhance stress tolerance traits. One of the early reports in engineering wax traits and thereby improved stress tolerance was from *Medicago sativa* (alfalfa), a forage legume. WXP1, a transcriptional regulator from *Medicago truncatula*, upregulated by drought, cold and ABA, was over expressed in alfalfa, which significantly increased the leaf cuticular wax load, mainly contributed by the C30 primary alcohol. The transgenic plants exhibited enhanced tolerance to drought and rapid recovery under rehydration [155]. Over expression of SlSHN1, a close homolog of the WIN/SHN gene from *Arabidopsis*, in tomato using constitutive CaMV 35S promoter improved drought tolerance, with higher cuticular wax deposition on leaf epidermal tissue. The transgenic plants displayed delayed wilting, improved water status and reduced water status [165]. MYB96, a transcriptional regulator over-expressed in *Camelina*, an emerging biofuel crop, which generated plants with enhanced drought tolerance. The expression levels of CsKCS2, CsKCS6, CsKCR1-1, CsKCR1-2, CsECR, and CsMAH1 were highly upregulated in the transgenic plants which resulted in a significant increase in the deposition of epicuticular waxes and total wax loads. This gives an option to cultivate the crop on marginal lands to produce renewable biofuels and bioresource [120]. It was further demonstrated that ectopic expression of DEWAX, a negative regulator of cuticular wax biosynthesis increased tolerance to *Botrytis cinerea* in *Camelina* [126]. A study from groundnut by over-expressing the KCS1 gene from a drought tolerant genotype improved cuticular was load and drought tolerance in a susceptible genotype [166]. Likewise, several of such regulators have been identified from model systems as well as crop species and used for engineering crop plants to enhance stress tolerance.

8. Options for manipulation of wax traits for individual and/combined stress tolerance

In crop plants, due to the nature of combined stressors interactions, the stress effect is not always additive [3]. While working with glossy mutants of *Zea mays* (*gl4*), an enhanced colonization of bacteria, was observed leading to more leaf blight pathogen growth compared to the wild type [167]. The thin cuticle provided leaf blight pathogen, an easy access to nutrient and water in *gl4* mutant indicating that cuticular wax thickness is a useful trait to identify plants' resistance to combined stressors. Additionally, wax layer structure and composition are equally important in conferring defense mechanisms. As rightly pointed in Ref. [1], such combined studies allow us to understand the shared and specific effects of biotic and abiotic stressors.

Wild relatives and landraces have long been recognized as a source of genes for breeding major field and horticulture crops. During domestication of wheat, tomato, rice, soybean and corn, yield was the focus trait. This in turn narrowed the genetic diversity for other biotic and abiotic stressors [168]. For example, during domestication of modern wheat, due to a phenotyping bottleneck a largely overlooked drought trait in wheat breeding program is glaucousness [169]. Such beneficial allelic variants lost in cuticle related traits can be introgressed back by crossing an elite line with its wild relatives. Apart from genetic diversity, a mutation population (EMS or gamma irradiation) provides an alternative avenue to target crop improvement via selection of cuticle-associated trait variations [170]. In fleshy tomatoes, a mutant line underlying for *delayed fruit deterioration* (DFD), is characterized for minimal transpirational water-loss and enhance post-harvest shelf life [171]. A recent alternative for trait manipulation is CRISPR-Cas9 system which is a precise gene-editing technology. This new method accelerates the evaluation of beneficial cuticle-associated alleles in different genetic backgrounds [172]. In similar lines, small RNA based transgenic strategy is also emerging as a molecule of choice to deal with combined biotic and abiotic resistance in crops [173].

9. Conclusion

There is sufficient evidence to argue that cuticle and cuticular waxes are involved in the regulation of multiple biotic and abiotic interactions. The cuticular wax can be treated as an important trait contributing for multiple stress resistance. Concerted efforts have been made to elucidate the synthesis and deposition of cuticular waxes in plants. Further analysis of the key regulatory steps involved in the formation of cuticular waxes, and also the role played by diverse types of wax components and structures in stress response is needed. This information could be incorporated in crop improvement programs (via marker assisted selection for wax genes). Since there are promising options emerging to analyze the cuticular wax trait using modern synchrotron technology [174] as well as now widely recognized techniques to observe ice propagation in real time across the cuticle [175] crop breeders have the potential to improve their efficiency of selection based on these traits. Recent progress in genomics can substantially help major field and horticulture crops to buffer the impacts of climate change. In addition, new genome-editing technologies will provide interesting tools to characterize and engineer waxes in crops. Unraveling key regulators and network partners of surface wax synthesis would aid in targeted manipulation of the trait using modern biotechnological applications. There are options to analyze the cuticular wax trait using modern non-destructive approaches. Crop breeders can use these tools to improve their efficiency of selection for the trait, and effectively pyramid the trait in elite genotypes to combat combined stresses.

Acknowledgements

RYS was supported by Agriculture and Agri-Food Canada. TR and KKT research was supported by the Agriculture Development Fund (Saskatchewan Ministry of Agriculture) and the Natural Science and Engineering Research Council (NSERC) Collaborative Research and Development program, Canada. NKN would like to acknowledge the Department of Biotechnology, Government of India, New Delhi (BT/TDS/121/SP20276/2016) and UAS Bengaluru (No. DR/Prof.(S)/RKVY/Alloc./B-44/2017-18) for the partial financial support.

Author details

Kunhikrishnan H. Dhanyalakshmi[1], Raju Y. Soolanayakanahally[2], Tawhidur Rahman[3], Karen K. Tanino[3] and Karaba N. Nataraja[1*]

1 Department of Crop Physiology, University of Agricultural Sciences, GKVK, Bengaluru, India

2 Saskatoon Research and Development Centre, Agriculture and Agri-Food Canada, Saskatoon, SK, Canada

3 Department of Plant Sciences, College of Agriculture and Bioresources, University of Saskatchewan, Saskatoon, SK, Canada

*Address all correspondence to: nataraja_karaba@yahoo.com

References

[1] Pandey P, Ramegowda V, Senthil-Kumar M. Shared and unique responses of plants to multiple individual stresses and stress combinations: Physiological and molecular mechanisms. Frontiers in Plant Science. 2015;**6**:723. DOI: 10.3389/fpls.2015.00723

[2] Kudela V. Potential impact of climate change on geographic distribution of plant pathogenic bacteria in Central Europe. Plant Protection Science. 2009;**45**:27-32. DOI: 10.17221/2832-PPS

[3] Atkinson NJ, Urwin PE. The interaction of plant biotic and abiotic stresses: From genes to the field. Journal of Experimental Botany. 2012;**63**:3523-3543. DOI: 10.1093/jxb/ers100

[4] Ramegowda V, Senthil-Kumar M. The interactive effects of simultaneous biotic and abiotic stresses on plants: Mechanistic understanding from drought and pathogen combination. Journal of Plant Physiology. 2015;**176**:47-54. DOI: 10.1016/j.jplph.2014.11.008

[5] Riederer M, Schreiber L. Protecting against water loss: Analysis of the barrier properties of plant cuticles. Journal of Experimental Botany. 2001;**52**:2023-2032. DOI: 10.1093/jexbot/52.363.2023

[6] Kerstiens G. Cuticular water permeability and its physiological significance. Journal of Experimental Botany. 1996;**47**:1813-1832. DOI: 10.1093/jxb/47.12.1813

[7] Kenrick P, Crane PR. The origin and early evolution of plants on land. Nature. 1997;**389**:33. DOI: 10.1038/37918

[8] Reina-Pinto JJ, Yephremov A. Surface lipids and plant defenses. Plant Physiology and Biochemistry. 2009;**47**:540-549. DOI: 10.1016/j.plaphy.2009.01.004

[9] Sieber P, Schorderet M, Ryser U, Buchala A, Kolattukudy P, Metraux JP, et al. Transgenic *Arabidopsis* plants expressing a fungal cutinase show alterations in the structure and properties of the cuticle and postgenital organ fusions. The Plant Cell. 2000;**12**:721-738. DOI: 10.1105/tpc.12.5.721

[10] Yeats TH, Rose JKC. The formation and function of plant cuticles. Plant Physiology. 2013;**163**:5-20. DOI: 10.1104/pp.113.222737

[11] Bellec Y, Harrar Y, Butaeye C, Darnet S, Bellini C, Faure JD. Pasticcino2 is a protein tyrosine phosphatase-like involved in cell proliferation and differentiation in *Arabidopsis*. The Plant Journal. 2002;**32**:713-722. DOI: 10.1046/j.1365-313X.2002.01456.x

[12] Wang Z-Y, Xiong L, Li W, Zhu J-K, Zhu J. The plant cuticle is required for osmotic stress regulation of abscisic acid biosynthesis and osmotic stress tolerance in *Arabidopsis*. The Plant Cell. 2011;**23**:1971-1984. DOI: 10.1105/tpc.110.081943

[13] Podila GK, Rogers LM, Kolattukudy PE. Chemical signals from avocado surface wax trigger germination and appressorium formation in *Colletotrichum gloeosporioides*. Plant Physiology. 1993;**103**:267-272. DOI: 10.1104/pp.103.1.267

[14] Martin LBB, Rose JKC. There's more than one way to skin a fruit: Formation and functions of fruit cuticles. Journal of Experimental Botany. 2014;**65**:4639-4651. DOI: 10.1093/jxb/eru301

[15] Kosma DK, Parsons EP, Isaacson T, Lu S, Rose JK, Jenks MA. Fruit cuticle lipid composition during development in tomato ripening mutants. Physiologia Plantarum. 2010;**139**:107-17. DOI: 10.1111/j.1399-3054.2009.01342.x

[16] Isabel L, Burcu B, Goulao Luis F. The fruit cuticle as a modulator of postharvest quality. Postharvest Biology and Technology. 2014;**87**:103-112. DOI: 10.1016/j.postharvbio.2013.08.012

[17] Isaacson T, Kosma DK, Matas AJ, Buda GJ, He Y, Yu B, et al. Cutin deficiency in the tomato fruit cuticle consistently affects resistance to microbial infection and biomechanical properties, but not transpirational water loss. The Plant Journal. 2009;**60**:363-77. DOI: 10.1111/j.1365-313X.2009.03969.x

[18] Baker EA. Chemistry and morphology of plant epicuticular waxes. In: Linnean Society Symposium Series. 1982

[19] Lee SB, Suh MC. Advances in the understanding of cuticular waxes in *Arabidopsis thaliana* and crop species. Plant Cell Reports. 2015;**34**:557-572. DOI: 10.1007/s00299-015-1772-2

[20] Schonherr J. Resistance of plant surfaces to water loss: Transport properties of cutin, suberin and associated lipids. In: Physiological Plant Ecology II. Berlin, Heidelberg: Springer; 1982. pp. 153-179

[21] Schönherr J, Riederer M. Foliar penetration and accumulation of organic chemicals in plant cuticles. In: Reviews of Environmental Contamination and Toxicology. New York, NY: Springer; 1989. pp. 1-70

[22] Jeffree CE. The fine structure of the plant cuticle. Biology of the Plant Cuticle. 2006;**23**:11-125. DOI: 10.1002/9780470988718.ch2

[23] Buschhaus C, Jetter R. Composition differences between epicuticular and intracuticular wax substructures: How do plants seal their epidermal surfaces? Journal of Experimental Botany. 2011;**62**:841-853. DOI: 10.1093/jxb/erq366

[24] Jetter R, Schäffer S. Chemical composition of the *Prunus laurocerasus* leaf surface. Dynamic changes of the epicuticular wax film during leaf development. Plant Physiology. 2001;**126**:1725-1737. DOI: 10.1104/pp.126.4.1725

[25] Jetter R, Kunst L, Samuels AL. Composition of plant cuticular waxes. Biology of the Plant Cuticle. 2008;**23**:145-181. DOI: 10.1002/9781119312994.apr0232

[26] Lee SB, Suh MC. Recent advances in cuticular wax biosynthesis and its regulation in *Arabidopsis*. Molecular Plant. 2013;**6**:246-249. DOI: 10.1093/mp/sss159

[27] Xue D, Zhang X, Lu X, Chen G, Chen Z-H. Molecular and evolutionary mechanisms of cuticular wax for plant drought tolerance. Frontiers in Plant Science. 2017;**8**:621. DOI: 10.3389/fpls.2017.00621

[28] Jetter R, Riederer M. Localization of the transpiration barrier in the epi-and intracuticular waxes of eight plant species: Water transport resistances are associated with fatty acyl rather than alicyclic components. Plant Physiology. 2016;**170**:921-934. DOI: 10.1104/pp.15.01699

[29] Zeisler-Diehl V, Müller Y, Schreiber L. Epicuticular wax on leaf cuticles does not establish the transpiration barrier, which is essentially formed by intracuticular wax. Journal of Plant Physiology. 2018;**227**:66-74. DOI: 10.1016/j.jplph.2018.03.018

[30] Zeisler V, Schreiber L. Epicuticular wax on cherry laurel (*Prunus laurocerasus*) leaves does not constitute the cuticular transpiration barrier. Planta. 2016;**243**:65-81. DOI: 10.1007/s00425-015-2397-y

[31] Mamrutha HM, Mogili T, Lakshmi KJ, Rama N, Kosma D, Kumar MU, et al. Leaf cuticular wax amount and crystal morphology regulate post-harvest

water loss in mulberry (*Morus* species). Plant Physiology and Biochemistry. 2010;**48**:690-696. DOI: 10.1016/j.plaphy.2010.04.007

[32] Shepherd T, Griffiths DW. The effects of stress on plant cuticular waxes. New Phytologist. 2006;**171**:469-499. DOI: 10.1111/j.1469-8137.2006.01826.x

[33] Seo PJ, Park C-M. Cuticular wax biosynthesis as a way of inducing drought resistance. Plant Signaling & Behavior. 2011;**6**:1043-1045. DOI: 10.4161/psb.6.7.15606

[34] Serrano M, Coluccia F, Torres M, L'Haridon F, Metraux JP. The cuticle and plant defense to pathogens. Frontiers in Plant Science. 2014;**5**:1-8. DOI: 10.3389/fpls.2014.00274

[35] Krauss P, Markstädter C, Riederer M. Attenuation of UV radiation by plant cuticles from woody species. Plant, Cell and Environment. 1997;**20**:1079-1085. DOI: 10.1111/j.1365-3040.1997.tb00684.x

[36] Solovchenko A, Merzlyak M. Optical properties and contribution of cuticle to UV protection in plants: Experiments with apple fruit. Photochemical & Photobiological Sciences. 2003;**2**:861-866. DOI: 10.1039/B302478D

[37] Zhu L, Guo J, Zhu J, Zhou C. Enhanced expression of EsWAX1 improves drought tolerance with increased accumulation of cuticular wax and ascorbic acid in transgenic *Arabidopsis*. Plant Physiology and Biochemistry. 2014;**75**:24-35. DOI: 10.1016/j.plaphy.2013.11.028

[38] Jordan WR, Shouse PJ, Blum A, Miller FR, Monk RL. Environmental physiology of sorghum. II. Epicuticular wax load and cuticular transpiration. Crop Science. 1984;**24**:1168-1173. DOI: 10.2135/cropsci1984.0011183X002400060038x

[39] Monneveux P, Reynolds MP, González-Santoyo H, Peña RJ, Mayr L, Zapata F. Relationships between grain yield, flag leaf morphology, carbon isotope discrimination and ash content in irrigated wheat. Journal of Agronomy and Crop Science. 2004;**190**:395-401. DOI: 10.1111/j.1439-037X.2004.00116.x

[40] Gonzalez A, Ayerbe L. Effect of terminal water stress on leaf epicuticular wax load, residual transpiration and grain yield in barley. Euphytica. 2010;**172**:341-349. DOI: 10.1007/s10681-009-0027-0

[41] Zhu X, Xiong L. Putative megaenzyme DWA1 plays essential roles in drought resistance by regulating stress-induced wax deposition in rice. Proceedings of the National Academy of Sciences. 2013;**110**:17790-17795. DOI: 10.1073/pnas.1316412110

[42] Sajeevan RS, Nataraja KN, Shivashankara KS, Pallavi N, Gurumurthy DS, Shivanna MB. Expression of Arabidopsis SHN1 in Indian mulberry (*Morus indica* L.) increases leaf surface wax content and reduces post-harvest water loss. Frontiers in Plant Science. 2017;**8**:418. DOI: 10.3389/fpls.2017.00418

[43] Hasanuzzaman M, Davies NW, Shabala L, Zhou M, Brodribb TJ, Shabala S. Residual transpiration as a component of salinity stress tolerance mechanism: A case study for barley. BMC Plant Biology. 2017;**17**:107. DOI: 10.1186/s12870-017-1054-y

[44] Tischler CR, Burson BL. Evaluating different bahiagrass cytotypes for heat tolerance and leaf epicuticular wax content. Euphytica. 1995;**84**:229-235. DOI: 10.1007/BF01681815

[45] Awika HO, Hays DB, Mullet JE, Rooney WL, Weers BD. QTL mapping and loci dissection for leaf epicuticular wax load and canopy temperature depression and their association with

QTL for staygreen in Sorghum bicolor under stress. Euphytica. 2017;**213**:207. DOI: 10.1007/s10681-017-1990-5

[46] Sanchez FJ, Manzanares M, de Andres EF, Tenorio JL, Ayerbe L. Residual transpiration rate, epicuticular wax load and leaf colour of pea plants in drought conditions. Influence on harvest index and canopy temperature. European Journal of Agronomy. 2001;**15**:57-70. DOI: 10.1016/S1161-0301(01)00094-6

[47] Heredia-Guerrero JA, Guzman-Puyol S, Benítez JJ, Athanassiou A, Heredia A, Domínguez E. Plant cuticle under global change: Biophysical implications. Global Change Biology. 2018;**24**:2749-51. DOI: 10.1111/gcb.14276

[48] Domínguez E, Heredia-Guerrero JA, Heredia A. The biophysical design of plant cuticles: An overview. New Phytologist. 2011;**189**:938-949. DOI: 10.1111/j.1469-8137.2010.03553.x

[49] Wisniewski M, Fuller M. Ice nucleation and deep supercooling in plants: New insights using infrared thermography BT-cold-adapted organisms: Ecology, physiology, enzymology and molecular biology. In: Margesin R, Schinner F, editors. Berlin, Heidelberg: Springer; 1999. pp. 105-118. DOI: 10.1007/978-3-662-06285-2_6

[50] Wisniewski M, Glenn DM, Fuller MP. Use of a hydrophobic particle film as a barrier to extrinsic ice nucleation in tomato plants. Journal of the American Society for Horticultural Science. 2002;**127**:358-364. DOI: 10.21273/JASHS.127.3.358

[51] Hamilton K. Identification of Ultrastructural and Biochemical Markers of Frost Avoidance in the Cuticular Layer of Corn. Canada: University of Saskatchewan; 2017

[52] Lindow SE, Arny DC, Upper CD. Erwinia herbicola: A bacterial ice nucleus active in increasing frost injury to corn. Phytopathology. 1978;**68**:523-527

[53] Lindow SE. The role of bacterial ICE nucleation in frost injury to plants. Annual Review of Phytopathology. 1983;**21**:363-384. DOI: 10.1146/annurev.py.21.090183.002051

[54] Storey K, Tanino K, editors. Nature at Risk: Temperature Adaptation and Climate Change. Wallingford, UK: CABI Press; 2011

[55] Gilbert RD, Johnson AM, Dean RA. Chemical signals responsible for appressorium formation in the rice blast fungus *Magnaporthe grisea*. Physiological and Molecular Plant Pathology. 1996;**48**:335-346. DOI: 10.1006/pmpp.1996.0027

[56] Ahmed A, Crawford T, Gould S, Ha YS, Hollrah M, Noor-E-Ain F, et al. Synthesis of (R)-and (S)-10, 16-dihydroxyhexadecanoic acid: Cutin stereochemistry and fungal activation. Phytochemistry. 2003;**63**:47-52. DOI: 10.1016/S0031-9422(03)00003-7

[57] Reisige K, Gorzelanny C, Daniels U, Moerschbacher BM. The C28 aldehyde octacosanal is a morphogenetically active component involved in host plant recognition and infection structure differentiation in the wheat stem rust fungus. Physiological and Molecular Plant Pathology. 2006;**68**:33-40. DOI: 10.1016/j.pmpp.2006.05.006

[58] Hegde Y, Kolattukudy PE. Cuticular waxes relieve self-inhibition of germination and appressorium formation by the conidia of *Magnaporthe grisea*. Physiological and Molecular Plant Pathology. 1997;**51**: 75-84. DOI: 10.1006/pmpp.1997.0105

[59] Hansjakob A, Riederer M, Hildebrandt U. Wax matters: Absence of very-long-chain aldehydes from the leaf cuticular wax of the glossy11 mutant of

maize compromises the prepenetration processes of *Blumeria graminis*. Plant Pathology. 2011;**60**:1151-1161. DOI: 10.1111/j.1365-3059.2011.02467.x

[60] Martin JT. Role of cuticle in the defense against plant disease. Annual Review of Phytopathology. 1964;**2**:81-100. DOI: 10.1146/annurev.py.02.090164.000501

[61] Barthlott W, Neinhuis C. Purity of the sacred lotus, or escape from contamination in biological surfaces. Planta. 1997;**202**:1-8. DOI: 10.1007/s004250050096

[62] Eigenbrode SD, Espelie KE. Effects of plant epicuticular lipids on insect herbivores. Annual Review of Entomology. 1995;**40**:171-194. DOI: 10.1146/annurev.en.40.010195.001131

[63] White C, Eigenbrode SD. Effects of surface wax variation in *Pisum sativum* on herbivorous and entomophagous insects in the field. Environmental Entomology. 2000;**29**:773-780. DOI: 10.1603/0046-225X-29.4.773

[64] Hariprasad K V, van Emden HF. Mechanisms of partial plant resistance to diamondback moth (*Plutella xylostella*) in brassicas. International Journal of Pest Management. 2010;**56**:15-22. DOI: 10.1080/09670870902980834

[65] Eigenbrode SD, Jetter R. Attachment to plant surface waxes by an insect predator. Integrative and Comparative Biology. 2002;**42**:1091-1099. DOI: 10.1093/icb/42.6.1091

[66] Wójcicka A. Surface waxes as a plant defense barrier towards grain aphid. Acta Biologica Cracoviensia s Botanica. 2015;**57**:95-103. DOI: 10.1515/abcsb-2015-0012

[67] Yeats TH, Buda GJ, Wang Z, Chehanovsky N, Moyle LC, Jetter R, et al. The fruit cuticles of wild tomato species exhibit architectural and chemical diversity, providing a new model for studying the evolution of cuticle function. The Plant Journal. 2012;**69**:655-666. DOI: 10.1111/j.1365-313X.2011.04820.x

[68] Kunst L, Samuels AL. Biosynthesis and secretion of plant cuticular wax. Progress in Lipid Research. 2003;**42**:51-80. DOI: 10.1016/S0163-7827(02)00045-0

[69] Post-Beittenmiller D. Biochemistry and molecular biology of wax production in plants. Annual Review of Plant Physiology and Plant Molecular Biology. 1996;**47**:405-430. DOI: 10.1146/annurev.arplant.47.1.405

[70] Beaudoin F, Wu X, Li F, Haslam RP, Markham JE, Zheng H, et al. Functional characterization of the *Arabidopsis* β-Ketoacyl-Coenzyme A reductase candidates of the fatty acid elongase. Plant Physiology. 2009;**150**:1174 LP-1191. DOI: 10.1104/pp.109.137497

[71] Kunst L, Samuels L. Plant cuticles shine: Advances in wax biosynthesis and export. Current Opinion in Plant Biology. 2009;**12**:721-727. DOI: 10.1016/j.pbi.2009.09.009

[72] McFarlane HE, Watanabe Y, Yang W, Huang Y, Ohlrogge J, Samuels AL. Golgi- and Trans-Golgi Network-mediated vesicle trafficking is required for wax secretion from epidermal cells. Plant Physiology. 2014;**164**:1250 LP-1260. DOI: 10.1104/pp.113.234583

[73] Bird D, Beisson F, Brigham A, Shin J, Greer S, Jetter R, et al. Characterization of *Arabidopsis* ABCG11/WBC11, an ATP binding cassette (ABC) transporter that is required for cuticular lipid secretion. The Plant Journal. 2007;**52**:485-498. DOI: 10.1111/j.1365-313X.2007.03252.x

[74] Debono A, Yeats TH, Rose JKC, Bird D, Jetter R, Kunst L, et al. *Arabidopsis* LTPG is a glycosylphosphatidylinositol-anchored

lipid transfer protein required for export of lipids to the plant surface. The Plant Cell. 2009;**21**:1230-1238. DOI: 10.1105/tpc.108.064451

[75] Lundqvist U, von Wettstein D. Induction of eceriferum mutants in barley by ionizing radiations and chemical mutagens. Hereditas. 1962;**48**:342-362. DOI: 10.1111/j.1601-5223.1962.tb01818.x

[76] Kurata T, Kawabata-Awai C, Sakuradani E, Shimizu S, Okada K, Wada T. The YORE-YORE gene regulates multiple aspects of epidermal cell differentiation in *Arabidopsis*. The Plant Journal. 2003;**36**:55-66. DOI: 10.1046/j.1365-313X.2003.01854.x

[77] Jenks MA, Tuttle HA, Eigenbrode SD, Feldmann KA. Leaf epicuticular waxes of the eceriferum mutants in *Arabidopsis*. Plant Physiology. 1995;**108**:369-377. DOI: 10.1104/pp.108.1.369

[78] Greer S, Wen M, Bird D, Wu X, Samuels L, Kunst L, et al. The Cytochrome P450 Enzyme CYP96A15 is the midchain alkane hydroxylase responsible for formation of secondary alcohols and ketones in stem cuticular wax of *Arabidopsis*. Plant Physiology. 2007;**145**:653 LP-667. DOI: 10.1104/pp.107.107300

[79] Lu S, Zhao H, Parsons EP, Xu C, Kosma DK, Xu X, et al. The glossyhead allele of ACC reveals a principal role for multidomain Acetyl-Coenzyme A Carboxylase in the biosynthesis of cuticular waxes by *Arabidopsis*. Plant Physiology. 2011;**157**:1079-1092. DOI: 10.1104/pp.111.185132

[80] Bonaventure G, Salas JJ, Pollard MR, Ohlrogge JB. Disruption of the FATB gene in *Arabidopsis* demonstrates an essential role of saturated fatty acids in plant growth. The Plant Cell. 2003;**15**:1020-1033. DOI: 10.1105/tpc.008946

[81] Fiebig A, Mayfield JA, Miley NL, Chau S, Fischer RL, Preuss D. Alterations in CER6, a gene identical to CUT1, differentially affect long-chain lipid content on the surface of pollen and stems. The Plant Cell. 2000;**12**:2001-2008. DOI: 10.1105/tpc.12.10.2001

[82] Sakuradani E, Zhao L, Haslam TM, Kunst L. The CER22 gene required for the synthesis of cuticular wax alkanes in *Arabidopsis thaliana* is allelic to CER1. Planta. 2013;**237**:731-738. DOI: 10.1007/s00425-012-1791-y

[83] Todd J, Post-Beittenmiller D, Jaworski JG. KCS1 encodes a fatty acid elongase 3-ketoacyl-CoA synthase affecting wax biosynthesis in *Arabidopsisthaliana*. The Plant Journal. 1999;**17**:119-130. DOI: 10.1046/j.1365-313X.1999.00352.x

[84] Lee S-B, Jung S-J, Go Y-S, Kim H-U, Kim J-K, Cho H-J, et al. Two *Arabidopsis* 3-ketoacyl CoA synthase genes, KCS20 and KCS2/DAISY, are functionally redundant in cuticular wax and root suberin biosynthesis, but differentially controlled by osmotic stress. The Plant Journal. 2009;**60**:462-475. DOI: 10.1111/j.1365-313X.2009.03973.x

[85] Lü S, Song T, Kosma DK, Parsons EP, Rowland O, Jenks MA. *Arabidopsis* CER8 encodes LONG-CHAIN ACYL-COA SYNTHETASE 1 (LACS1) that has overlapping functions with LACS2 in plant wax and cutin synthesis. The Plant Journal. 2009;**59**:553-564. DOI: 10.1111/j.1365-313X.2009.03892.x

[86] Kim J, Jung JH, Lee SB, Go YS, Kim HJ, Cahoon R, et al. *Arabidopsis* 3-Ketoacyl-Coenzyme A Synthase9 is involved in the synthesis of tetracosanoic acids as precursors of cuticular waxes, suberins,

sphingolipids, and phospholipids. Plant Physiology. 2013;**162**:567-580. DOI: 10.1104/pp.112.210450

[87] Chen X, Goodwin SM, Boroff VL, Liu X, Jenks MA. Cloning and characterization of the WAX2 gene of *Arabidopsis* involved in cuticle membrane and wax production. The Plant Cell. 2003;**15**:1170-1185. DOI: 10.1105/tpc.010926

[88] Zheng H, Rowland O, Kunst L. Disruptions of the *Arabidopsis* Enoyl-CoA reductase gene reveal an essential role for very-long-chain fatty acid synthesis in cell expansion during plant morphogenesis. The Plant Cell. 2005;**17**:1467-1481. DOI: 10.1105/tpc.104.030155

[89] Rowland O, Zheng H, Hepworth SR, Lam P, Jetter R, Kunst L. CER4 encodes an alcohol-forming fatty acyl-coenzyme A reductase involved in cuticular wax production in *Arabidopsis*. Plant Physiology. 2006;**142**:866-877. DOI: 10.1104/pp.106.086785

[90] Li F, Wu X, Lam P, Bird D, Zheng H, Samuels L, et al. Identification of the wax ester Synthase/Acyl-Coenzyme A:Diacylglycerol Acyltransferase WSD1 required for stem wax ester biosynthesis in *Arabidopsis*. Plant Physiology. 2008;**148**:97-107. DOI: 10.1104/pp.108.123471

[91] Bach L, Michaelson L V, Haslam R, Bellec Y, Gissot L, Marion J, et al. The very-long-chain hydroxy fatty acyl-CoA dehydratase PASTICCINO2 is essential and limiting for plant development. Proceedings of the National Academy of Sciences. 2008;**105**:14727-14731. DOI: 10.1073/pnas.0805089105

[92] Haslam TM, Mañas-Fernández A, Zhao L, Kunst L. *Arabidopsis* ECERIFERUM2 is a component of the fatty acid elongation machinery required for fatty acid extension to exceptional lengths. Plant Physiology. 2012;**160**:1164 -1174. DOI: 10.1104/pp.112.201640

[93] Yang X, Zhao H, Kosma DK, Tomasi P, Dyer JM, Li R, et al. The acyl desaturase CER17 is involved in producing wax unsaturated primary alcohols and cutin monomers. Plant Physiology. 2017;**173**:1109-1124. DOI: 10.1104/pp.16.01956

[94] Pighin JA, Zheng H, Balakshin LJ, Goodman IP, Western TL, Jetter R, et al. Plant cuticular lipid export requires an ABC transporter. Science. 2004;**306**:702-704. DOI: 10.1126/science

[95] Luo B, Xue X-Y, Hu W-L, Wang L-J, Chen X-Y. An ABC transporter gene of Arabidopsis thaliana, AtWBC11, is involved in cuticle development and prevention of organ fusion. Plant and Cell Physiology. 2007;**48**:1790-802. DOI: 10.1093/pcp/pcm152

[96] Kim H, Lee SB, Kim HJ, Min MK, Hwang I, Suh MC. Characterization of Glycosylphosphatidylinositol-Anchored Lipid Transfer Protein 2 (LTPG2) and overlapping function between LTPG/LTPG1 and LTPG2 in cuticular wax export or accumulation in *Arabidopsis thaliana*. Plant and Cell Physiology. 2012;**53**:1391-1403. DOI: 10.1093/pcp/pcs083

[97] Nobusawa T, Okushima Y, Nagata N, Kojima M, Sakakibara H, Umeda M. Synthesis of very-long-chain fatty acids in the epidermis controls plant organ growth by restricting cell proliferation. PLoS Biology. 2013;**11**:1-14. DOI: 10.1371/journal.pbio.1001531

[98] Rhee Y, Hlousek-Radojcic A, Ponsamuel J, Liu D, Post-Beittenmiller D. Epicuticular wax accumulation and fatty acid elongation activities are induced during leaf development of leeks. Plant Physiology. 1998;**116**:901 LP-911. DOI: 10.1104/pp.116.3.901

[99] Kamigaki A, Kondo M, Mano S, Hayashi M, Nishimura M. Suppression of peroxisome biogenesis factor 10 reduces cuticular wax accumulation by disrupting the er network in *Arabidopsis thaliana*. Plant and Cell Physiology. 2009;**50**:2034-2046. DOI: 10.1093/pcp/pcp152

[100] Mamrutha HM, Nataraja KN, Rama N, Kosma DK, Mogili T, Lakshmi KJ, et al. Leaf surface wax composition of genetically diverse mulberry (*Morus* sp.) genotypes and its close association with expression of genes involved in wax metabolism. Current Science. 2017;**112**:759-766. DOI: 10.18520/cs/v112/i04/759-766

[101] Steinmüller D, Tevini M. Action of ultraviolet radiation (UV-B) upon cuticular waxes in some crop plants. Planta. 1985;**164**:557-564. DOI: 10.1007/BF00395975

[102] Kakani VG, Reddy KR, Zhao D, Sailaja K. Field crop responses to ultraviolet-B radiation: A review. Agricultural and Forest Meteorology. 2003;**120**:191-218. DOI: 10.1016/j.agrformet.2003.08.015

[103] Shankar R, Bhattacharjee A, Jain M. Transcriptome analysis in different rice cultivars provides novel insights into desiccation and salinity stress responses. Scientific Reports. 2016;**6**:23719. DOI: 10.1038/srep23719

[104] Islam MA, Du H, Ning J, Ye H, Xiong L. Characterization of glossy1-homologous genes in rice involved in leaf wax accumulation and drought resistance. Plant Molecular Biology. 2009;**70**:443-56. DOI: 10.1007/s11103-009-9483-0

[105] Guo J, Xu W, Yu X, Shen H, Li H, Cheng D, et al. Cuticular wax accumulation is associated with drought tolerance in wheat near-isogenic lines. Frontiers in Plant Science. 2016;**7**:1809. DOI: 10.3389/fpls.2016.01809

[106] Bengtson C, Larsson S, Liljenberg C. Effects of water stress on cuticular transpiration rate and amount and composition of epicuticular wax in seedlings of six oat varieties. Physiologia Plantarum. 1978;**44**:319-324. DOI: 10.1111/j.1399-3054.1978.tb01630.x

[107] Saneoka H, Ogata S. Relationship between water use efficiency and cuticular wax deposition in warm season forage crops grown under water deficit conditions. Soil Science and Plant Nutrition. 1987;**33**:439-448. DOI: 10.1080/00380768.1987.10557590

[108] Wang Y, Wang M, Sun Y, Wang Y, Li T, Chai G, et al. FAR5, a fatty acyl-coenzyme A reductase, is involved in primary alcohol biosynthesis of the leaf blade cuticular wax in wheat (*Triticum aestivum* L.). Journal of Experimental Botany. 2015;**66**:1165-78. DOI: 10.1093/jxb/eru457

[109] Bourdenx B, Bernard A, Domergue F, Pascal S, Léger A, Roby D, et al. Overexpression of *Arabidopsis* ECERIFERUM1Promotes wax very-long-chain alkane biosynthesis and influences plant response to biotic and abiotic stresses. Plant Physiology. 2011;**156**: 29-45. DOI: 10.1104/pp.111.172320

[110] Kim H, Choi D, Suh MC. Cuticle ultrastructure, cuticular lipid composition, and gene expression in hypoxia-stressed *Arabidopsis* stems and leaves. Plant Cell Reports. 2017;**36**: 815-827. DOI: 10.1007/s00299-017-2112-5

[111] Amid A, Lytovchenko A, Fernie AR, Warren G, Thorlby GJ. The sensitive to freezing3 mutation of *Arabidopsis thaliana* is a cold-sensitive allele of homomeric acetyl-CoA carboxylase that results in cold-induced cuticle deficiencies. Journal of Experimental Botany. 2012;**63**:5289-5299. DOI: 10.1093/jxb/ers191

[112] Gray JE, Holroyd GH, van der Lee FM, Bahrami AR, Sijmons PC,

Woodward FI, et al. The HIC signalling pathway links CO_2 perception to stomatal development. Nature. 2000;**408**:713. DOI: 10.1038/35047071

[113] Park CS, Go YS, Suh MC. Cuticular wax biosynthesis is positively regulated by WRINKLED4, an AP2/ERF-type transcription factor, in *Arabidopsis* stems. The Plant Journal. 2016;**88**: 257-270. DOI: 10.1111/tpj.13248

[114] Broun P, Poindexter P, Osborne E, Jiang C-Z, Riechmann JL. WIN1, a transcriptional activator of epidermal wax accumulation in *Arabidopsis*. Proceedings of the National Academy of Sciences of the United States of America. 2004;**101**:4706-4711. DOI: 10.1073/pnas.0305574101

[115] Aharoni A, Dixit S, Jetter R, Thoenes E, van Arkel G, Pereira A. The SHINE Clade of AP2 domain transcription factors activates wax biosynthesis, alters cuticle properties, and confers drought tolerance when overexpressed in *Arabidopsis*. The Plant Cell. 2004;**16**:2463-2480. DOI: 10.1105/tpc.104.022897

[116] Wang Y, Wan L, Zhang L, Zhang Z, Zhang H, Quan R, et al. An ethylene response factor OsWR1 responsive to drought stress transcriptionally activates wax synthesis related genes and increases wax production in rice. Plant Molecular Biology. 2012;**78**: 275-288. DOI: 10.1007/s11103-011-9861-2

[117] Bi H, Shi J, Kovalchuk N, Luang S, Bazanova N, Chirkova L, et al. Overexpression of the TaSHN1 transcription factor in bread wheat leads to leaf surface modifications, improved drought tolerance, and no yield penalty under controlled growth conditions. Plant, Cell and Environment. 2018;**41**:2549-2566. DOI: 10.1111/pce.13339

[118] Go YS, Kim H, Kim HJ, Suh MC. Arabidopsis cuticular wax biosynthesis is negatively regulated by the DEWAX gene encoding an AP2/ERF-Type transcription factor. The Plant Cell. 2014;**26**:1666-1680. DOI: 10.1105/tpc.114.123307

[119] Kim H, Go YS, Suh MC. DEWAX2 transcription factor negatively regulates cuticular wax biosynthesis in *Arabidopsis* leaves. Plant and Cell Physiology. 2018;**59**:966-977. DOI: 10.1093/pcp/pcy033

[120] Lee SB, Kim H, Kim RJ, Suh MC. Overexpression of *Arabidopsis* MYB96 confers drought resistance in Camelina sativa via cuticular wax accumulation. Plant Cell Reports. 2014;**33**:1535-1546. DOI: 10.1007/s00299-014-1636-1

[121] Seo PJ, Lee SB, Suh MC, Park M-J, Go YS, Park C-M. The MYB96 transcription factor regulates cuticular wax biosynthesis under drought conditions in *Arabidopsis*. The Plant Cell. 2011;**23**:1138-1152. DOI: 10.1105/tpc.111.083485

[122] Seo PJ, Park CM. MYB96-mediated abscisic acid signals induce pathogen resistance response by promoting salicylic acid biosynthesis in *Arabidopsis*. New Phytologist. 2010;**186**:471-483. DOI: 10.1111/j.1469-8137.2010.03183.x

[123] Lee SB, Kim HU, Suh MC. MYB94 and MYB96 additively activate cuticular wax biosynthesis in *Arabidopsis*. Plant and Cell Physiology. 2016;**57**:2300-2311. DOI: 10.1093/pcp/pcw147

[124] Hooker TS, Lam P, Zheng H, Kunst L. A core subunit of the RNA-processing/degrading exosome specifically influences cuticular wax biosynthesis in *Arabidopsis*. The Plant Cell. 2007;**19**:904-913. DOI: 10.1105/tpc.106.049304

[125] Wang W, Liu X, Gai X, Ren J, Liu X, Cai Y, et al. *Cucumis sativus* L. WAX2 plays a pivotal role in wax biosynthesis, influencing pollen fertility and plant

biotic and abiotic stress responses. Plant and Cell Physiology. 2015;**56**:1339-1354. DOI: 10.1093/pcp/pcv052

[126] Ju S, Go YS, Choi HJ, Park JM, Suh MC. DEWAX transcription factor is involved in resistance to *Botrytis cinerea* in *Arabidopsis thaliana* and *Camelina sativa*. Frontiers in Plant Science. 2017;**8**:1210. DOI: 10.3389/fpls.2017.01210

[127] Ziv C, Zhao Z, Gao YG, Xia Y. Multifunctional roles of plant cuticle during plant-pathogen interactions. Frontiers in Plant Science. 2018;**9**:1088. DOI: 10.3389/fpls.2018.01088

[128] Woloshuk CP, Kolattukudy PE. Mechanism by which contact with plant cuticle triggers cutinase gene expression in the spores of *Fusarium solani* f. sp. pisi. Proceedings of the National Academy of Sciences. 1986;**83**: 1704-1708. DOI: 10.1073/pnas.83.6.1704

[129] Leroch M, Kleber A, Silva E, Coenen T, Koppenhöfer D, Shmaryahu A, et al. Transcriptome profiling of Botrytis cinerea conidial germination reveals upregulation of infection-related genes during the prepenetration stage. Eukaryotic Cell. 2013;**12**:614-626. DOI: 10.1128/EC.00295-12

[130] Garrido SM, Kitamoto N, Watanabe A, Shintani T, Gomi K. Functional analysis of FarA transcription factor in the regulation of the genes encoding lipolytic enzymes and hydrophobic surface binding protein for the degradation of biodegradable plastics in *Aspergillus oryzae*. Journal of Bioscience and Bioengineering. 2012;**113**:549-555. DOI: 10.1016/j.jbiosc.2011.12.014

[131] Wang B, Liang X, Gleason ML, Zhang R, Sun G. Genome sequence of the ectophytic fungus Ramichloridium luteum reveals unique evolutionary adaptations to plant surface niche. BMC Genomics. 2017;**18**:729. DOI: 10.1186/s12864-017-4118-3

[132] Malinovsky FG, Fangel JU, Willats WGT. The role of the cell wall in plant immunity. Frontiers in Plant Science. 2014:**5**:178. DOI: 10.3389/fpls.2014.00178

[133] Marques JPR, Amorim L, Spósito MB, Appezzato-da-Glória B. Ultrastructural changes in the epidermis of petals of the sweet orange infected by *Colletotrichum acutatum*. Protoplasma. 2016;**253**:1233-1242. DOI: 10.1007/s00709-015-0877-3

[134] Alkan N, Fortes AM. Insights into molecular and metabolic events associated with fruit response to post-harvest fungal pathogens. Frontiers in Plant Science. 2015;**6**:889. DOI: 10.3389/fpls.2015.00889

[135] Xia Y, Gao Q-M, Yu K, Lapchyk L, Navarre D, Hildebrand D, et al. An intact cuticle in distal tissues is essential for the induction of systemic acquired resistance in plants. Cell Host and Microbe. 2009;**5**:151-165. DOI: 10.1016/j.chom.2009.01.001

[136] Xia Y, Yu K, Navarre D, Seebold K, Kachroo A, Kachroo P. The glabra1 mutation affects cuticle formation and plant responses to microbes. Plant Physiology. 2010;**154**:833-846. DOI: 10.1104/pp.110.161646

[137] Sela D, Buxdorf K, Shi JX, Feldmesser E, Schreiber L, Aharoni A, et al. Overexpression of AtSHN1/WIN1 provokes unique defense responses. PLoS One. 2013;**8**:e70146. DOI: 10.1371/journal.pone.0070146

[138] Buxdorf K, Rubinsky G, Barda O, Burdman S, Aharoni A, Levy M. The transcription factor SlSHINE3 modulates defense responses in tomato plants. Plant Molecular Biology. 2014;**84**:37-47. DOI: 10.1007/s11103-013-0117-1

[139] Koch K, Ensikat H-J. The hydrophobic coatings of plant surfaces:

Epicuticular wax crystals and their morphologies, crystallinity and molecular self-assembly. Micron. 2008;**39**:759-772. DOI: 10.1016/j.micron.2007.11.010

[140] Bargel H, Koch K, Cerman Z, Neinhuis C. Evans Review No. 3: Structurefunction relationships of the plant cuticle and cuticular waxes a smart material? Functional Plant Biology. 2006;**33**:893-910. DOI: 10.1071/FP06139

[141] Gorb E V, Gorb SN. Anti-adhesive effects of plant wax coverage on insect attachment. Journal of Experimental Botany. 2017;**68**:5323-5337. DOI: 10.1093/jxb/erx271

[142] Brennan EB, Weinbaum SA. Effect of epicuticular wax on adhesion of psyllids to glaucous juvenile and glossy adult leaves of *Eucalyptus globulus* Labillardière. Australian Journal of Entomology. 2001;**40**:270-277. DOI: 10.1046/j.1440-6055.2001.00229.x

[143] Bodnaryk RP. Leaf epicuticular wax, an antixenotic factor in *Brassicaceae* that affects the rate and pattern of feeding of flea beetles, *Phyllotreta cruciferae* (Goeze). Canadian Journal of Plant Science. 1992;**72**: 1295-1303. DOI: 10.4141/cjps92-163

[144] Eigenbrode SD. The effects of plant epicuticular waxy blooms on attachment and effectiveness of predatory insects. Arthropod Structure and Development. 2004;**33**:91-102. DOI: 10.1016/j.asd.2003.11.004

[145] Eigenbrode SD, White C, Rohde M, Simon CJ. Behavior and effectiveness of adult *Hippodamia convergens* (Coleoptera: Coccinellidae) as a predator of *Acyrthosiphon pisum* (Homoptera: Aphididae) on a wax mutant of *Pisum sativum*. Environmental Entomology. 1998;**27**:902-909. DOI: 10.1093/ee/27.4.902

[146] Gorb E, Gorb S. How a lack of choice can force ants to climb up waxy plant stems. Arthropod-Plant Interactions. 2011;**5**:297-306. DOI: 10.1007/s11829-011-9143-6

[147] Gorb E, Haas K, Henrich A, Enders S, Barbakadze N, Gorb S. Composite structure of the crystalline epicuticular wax layer of the slippery zone in the pitchers of the carnivorous plant *Nepenthes alata* and its effect on insect attachment. Journal of Experimental Biology. 2005;**208**:4651-4662. DOI: 10.1242/jeb.01939

[148] Gorb E, Böhm S, Jacky N, Maier L-P, Dening K, Pechook S, et al. Insect attachment on crystalline bioinspired wax surfaces formed by alkanes of varying chain lengths. Beilstein Journal of Nanotechnology. 2014;**5**:1031-1041. DOI: 10.3762/bjnano.5.116

[149] Borodich FM, Gorb E V, Gorb SN. Fracture behaviour of plant epicuticular wax crystals and its role in preventing insect attachment: A theoretical approach. Applied Physics A. 2010;**100**:63-71. DOI: 10.1007/s00339-010-5794-x

[150] Dirks J-H, Clemente CJ, Federle W. Insect tricks: Two-phasic foot pad secretion prevents slipping. Journal of the Royal Society, Interface. 2010;**7**: 587-593. DOI: 10.1098/rsif.2009.0308

[151] Gorb E V, Hofmann P, Filippov AE, Gorb SN. Oil adsorption ability of three-dimensional epicuticular wax coverages in plants. Scientific Reports. 2017;**7**:45483. DOI: 10.1038/srep45483

[152] Edwards D, Abbot GD, Raven JA. Cuticles of early land plants: A palaeoecophysiological evaluation. In: Kersteins G, editor. Plant Cuticles: An Integrated Functional Approach. Oxford: Bios; 1996. pp. 1-31

[153] Qin B-X, Tang D, Huang J, Li M, Wu X-R, Lu L-L, et al. Rice OsGL1-1 is

involved in leaf cuticular wax and cuticle membrane. Molecular Plant. 2011;**4**: 985-995. DOI: 10.1093/mp/ssr028

[154] Cameron KD, Teece MA, Smart LB. Increased accumulation of cuticular wax and expression of lipid transfer protein in response to periodic drying events in leaves of tree tobacco. Plant Physiology. 2006;**140**:176-183. DOI: 10.1104/pp.105.069724

[155] Zhang J-Y, Broeckling CD, Blancaflor EB, Sledge MK, Sumner LW, Wang Z-Y. Overexpression of WXP1, a putative *Medicago truncatula* AP2 domain-containing transcription factor gene, increases cuticular wax accumulation and enhances drought tolerance in transgenic alfalfa (*Medicago sativa*). The Plant Journal. 2005;**42**:689-707. DOI: 10.1111/j.1365-313X.2005.02405.x

[156] Febrero A, Fernández S, Molina-Cano JL, Araus JL. Yield, carbon isotope discrimination, canopy reflectance and cuticular conductance of barley isolines of differing glaucousness. Journal of Experimental Botany. 1998;**49**: 1575-1581. DOI: 10.1093/jxb/49.326.1575

[157] Richards RA, Rawson HM, Johnson DA. Glaucousness in wheat: Its development and effect on water-use efficiency, gas exchange and photosynthetic tissue temperatures. Functional Plant Biology. 1986;**13**: 465-473. DOI: 10.1071/PP9860465

[158] Johnson DA, Richards RA, Turner NC. Yield, water relations, gas exchange, and surface reflectances of near-isogenic wheat lines differing in glaucousness. Crop Science. 1983;**23**:318-325. DOI: 10.2135/cropsci1983.0011183X002300020033x

[159] Jansen MAK, Gaba V, Greenberg BM. Higher plants and UV-B radiation: Balancing damage, repair and acclimation. Trends in Plant Science. 1998;**3**:131-135. DOI: 10.1016/S1360-1385(98)01215-1

[160] Fukuda S, Satoh A, Kasahara H, Matsuyama H, Takeuchi Y. Effects of ultraviolet-B irradiation on the cuticular wax of cucumber (*Cucumis sativus*) cotyledons. Journal of Plant Research. 2008;**121**:179-189. DOI: 10.1007/s10265-007-0143-7

[161] Srinivasan S, Gomez SM, Kumar SS, Ganesh SK, Biji KR, Senthil A, et al. QTLs linked to leaf epicuticular wax, physio-morphological and plant production traits under drought stress in rice (*Oryza sativa* L.). Plant Growth Regulation. 2008;**56**:245-256. DOI: 10.1007/s10725-008-9304-5

[162] Burow GB, Franks CD, Acosta-Martinez V, Xin Z. Molecular mapping and characterization of BLMC, a locus for profuse wax (bloom) and enhanced cuticular features of Sorghum (*Sorghum bicolor* (L.) Moench.). Theoretical and Applied Genetics. 2009;**118**:423-431. DOI: 10.1007/s00122-008-0908-y

[163] Liu D, Dong X, Liu Z, Tang J, Zhuang M, Zhang Y, et al. Fine mapping and candidate gene identification for wax biosynthesis locus, BoWax1 in *Brassica oleracea* L. var. capitata. Frontiers in Plant Science. 2018:**9**:309. DOI: 10.3389/fpls.2018.00309

[164] Debieu M, Sine B, Passot S, Grondin A, Akata E, Gangashetty P, et al. Response to early drought stress and identification of QTLs controlling biomass production under drought in pearl millet. PLoS One. 2018;**13**:e0201635. DOI: 10.1371/journal.pone.0201635

[165] Al-Abdallat AM, Al-Debei HS, Ayad JY, Hasan S. Over-expression of SlSHN1 gene improves drought tolerance by increasing cuticular wax accumulation in tomato. International Journal of Molecular Sciences. 2014;**15**:19499-19515. DOI: 10.3390/ijms151119499

[166] Lokesh U, Venkatesh B, Kiranmai K, Nareshkumar A, Amarnathareddy V, Rao GL, et al. Overexpression of

ß-Ketoacyl Co-A Synthase1 gene improves tolerance of drought susceptible groundnut (*Arachis hypogaea* L.) cultivar K-6 by increased leaf epicuticular wax accumulation. Frontiers in Plant Science. 2019;**9**:1869. DOI: 10.3389/fpls.2018.01869

[167] Marcell LM, Beattie GA. Effect of leaf surface waxes on leaf colonization by*Pantoea agglomerans* and *Clavibacter michiganensis*. Molecular Plant-Microbe Interactions. 2002;**15**:1236-1244. DOI: 10.1094/MPMI.2002.15.12.1236

[168] Tanksley SD, McCouch SR. Seed banks and molecular maps: Unlocking genetic potential from the wild. Science. 1997;**277**:1063-1066. DOI: 10.1126/science.277.5329.1063

[169] Hen-Avivi S, Savin O, Racovita RC, Lee W-S, Adamski NM, Malitsky S, et al. A Metabolic gene cluster in the wheat W1 and the barley Cer-cqu loci determines β-Diketone biosynthesis and glaucousness. The Plant Cell. 2016;**28**:1440-1460. DOI: 10.1105/tpc.16.00197

[170] Oladosu Y, Rafii MY, Abdullah N, Hussin G, Ramli A, Rahim HA, et al. Principle and application of plant mutagenesis in crop improvement: A review. Biotechnology and Biotechnological Equipment. 2016;**30**:1-16. DOI: 10.1080/13102818.2015.1087333

[171] Saladié M, Matas AJ, Isaacson T, Jenks MA, Goodwin SM, Niklas KJ, et al. A reevaluation of the key factors that influence tomato fruit softening and integrity. Plant Physiology. 2007;**144**:1012-1028. DOI: 10.1104/pp.107.097477

[172] Cermak T, Curtin SJ, Gil-Humanes J, Cegan R, Kono TJY, Konecna E, et al. A multipurpose toolkit to enable advanced genome engineering in plants. The Plant Cell. 2017;**29**:1196-1217. DOI: 10.1105/tpc.16.00922

[173] Sajeevan RS, Parvathi MS, Nataraja KN. Leaf wax trait in crops for drought and biotic stress tolerance: Regulators of epicuticular wax synthesis and role of small RNAs. Indian Journal of Plant Physiology. 2017;**22**:434-447. DOI: 10.1007/s40502-017-0333-9

[174] Willick IR, Lahlali R, Vijayan P, Muir D, Karunakaran C, Tanino KK. Wheat flag leaf epicuticular wax morphology and composition in response to moderate drought stress are revealed by SEM, FTIR-ATR and synchrotron X-ray spectroscopy. Physiologia Plantarum. 2018;**162**:316-332. DOI: 10.1111/ppl.12637

[175] Wisniewski M, Lindow SE, Ashworth EN. Observations of ice nucleation and propagation in plants using infrared video thermography. Plant Physiology. 1997;**113**:327-334. DOI: 10.1104/pp.113.2.327

Section 2

Biotic Stress

Chapter 4

Fungal Biotic Stresses in Plants and Its Control Strategy

Faisal Hussain and Farzana Usman

Abstract

Vegetable and fruit are economically very important and valuable crops throughout the world. According to Pakistan Agriculture Research Council (PARC) the same report, Pakistan was the fifth largest exporter in the world, but according to there is continuous decline in the production of fruit and vegetable in Pakistan. The reasons for this reduction are various and many but the major threat in yield production is various pests and pathogens which cause considerable losses every year. Major insects which attack on crop fields are aphids, mites, thrips, etc.; besides pests, different pathogens also cause various diseases in field crop and reduce yield of the plant, for example, fungi, viruses, bacteria, and nematodes. Among other pathogens, the fungal diseases are more destructive than diseases cause by other pathogens.

Keywords: fungi, phytopathogens, vegetables, losses, yield

1. Introduction

Commonly fungal diseases cause more damage than diseases cause by other pathogens. Several fungi cause various diseases in different crop fields and plants (**Table 1**). Sometime these fungi can cause similar symptoms and confused to

Crop stage	Common fungal diseases	Causal agent
Vegetative stage		
Seedling	Damping off, *Phytophthora* blight, *Fusarium* wilt	*Pythium* spp., *Phytophthora* sp., *Fusarium oxysporum*
After transplantation	Frog eye leaf spot, damping off, *Fusarium* wilt	*Cercospora capsici*, *Pythium* spp., *Fusarium oxysporum*
Root diseases	*Phytophthora* root rot, *Verticillium* wilt, *Rhizoctonia* root rot	*Phytophthora capsici*, *Verticillium* sp., *Rhizoctonia solani*
Reproductive stage		
Flowering	Powdery mildew	*Leveillula taurica*
Fruiting	*Aspergillus*, anthracnose, powdery mildew, *Phytophthora* pod rot, black mold	*Aspergillus* spp., *Colletotrichum* sp., *Leveillula taurica*, *Phytophthora capsici*, *Alternaria* spp.

Table 1.
Fungal diseases and their causal agents attacking on chili plant at different crop stages.

Regions	Subregions	Major areas in Pakistan
Saharo-Sindian		Lower Sindh, some areas of upper Sindh, Central and Southern Punjab, Southern Balochistan and KPK plains, Rawalpindi, Islamabad
Irano-Turanian	a. Western subregions	Waziristan and North Baluchistan Upper portion of northern areas, Baltistan, Chitral
	b. Eastern subregions or Central Asia	
Sino-Japanese		Kashmir, KPK, Der, Swat, Muzaffarabad, Hazaras
Indian		East and west of river Jhelum, Lahore

Table 2.
Major chili-producing phytogeographical regions of Pakistan.

one another. There are some preharvesting fungal diseases such as *Aspergillus*, *Alternaria*, anthracnose, *Cercospora* (frogeye) leaf spot, charcoal rot, *Choanephora* blight (wet rot), damping-off, root-rot, downy mildew, *Fusarium* stem rot, *Fusarium* wilt, gray leaf spot, gray mold, *Phytophthora* blight, powdery mildew, Southern blight, *Verticillium* wilt, and white mold that are grown in different crops.

The vegetable and cereal crop are recorded about the entire region and areas of Pakistan. So there are no clear-cut boundaries between each region and intermingling of crop filed occurs. Pakistan has four major phytogeographical regions [1, 2]. Major floristic zones are summarized in **Table 2**.

2. Major fungal diseases causing loss of crops

Commonly fungal diseases cause more damage than diseases caused by other pathogens. Several fungi caused various diseases in field crops. Some major fungal diseases which are responsible for heavy loss to the field crops are as follows:

2.1 *Alternaria* spp.

Genus *Alternaria* is considered as seed-borne fungi, widespread, and highly cause of decline that usually infects chili plants. The loss of fruit yield has been recorded to be up to 100% under certain environment conditions [3]. The disease is characterized by fairly firm, superficial lesions which often occur on berries near the pedicel. Lesions are tan at first, become dark brown to black with age, and remain localized. Under humid conditions provided by cold transit, fluffy gray tufts of fungus often occur on rachis and pedicels, occasionally without causing visible lesions [4]. Symptoms of *Alternaria* begin as water-soaked, gray lesion on either the side or blossom-end of the fruit [5]. As the lesion progresses, they darken and become covered with spores. Internal necrosis and mycelial growth occur on the seeds, placenta, and pericarp but are not noticed until the harvesting [6]. Infection can occur through the flowers or insect injury, mechanical damage, chilling injury, sunburn, or blossom-end rot [6–9].

2.2 Anthracnose/dieback/fruit rot (*Colletotrichum* spp.)

Anthracnose or dieback and fruit rot are caused by *Colletotrichum* sp. [10–12]. The fruit-rot diseases the diseases of crops which belong to the tropical regions has been reported heavy damage. It is remarked as major harmful disease of vegetable crop [13]. Anthracnose also caused by *Colletotrichum* spp. is a major problem of ripened fruit (Hence, it is also called as ripe fruit rot.), occurs worldwide wherever cereal and other crops are grown [14], is severely infected by anthracnose which may cause yield losses up to 50% [15]. It is a seed-borne fungus. It produces dark spot, water-soaked lesions that rapidly expand. In some cases, the lesions are brown and then turn black from the formation of setae and sclerotia [16]. Infections spread rapidly during periods of excess irrigation or rain on immature pods.

2.3 *Aspergillus* spp.

Several fungi may be present on fruit after the postharvest and storage due to the relative humidity is not controlled during the storage process [17]. There are a number of *Aspergillus* sp. recorded in different crop-growing areas of Pakistan including *Aspergillus flavus*, *A. niger*, *A. fumigates*, etc. Several fruit and vegetables are reported to be contaminated with molds and their toxic metabolites, and *Aspergillus flavus* is the predominant mold on collected samples in several cases [18–22]. Aflatoxins are chemically classified in secondary metabolite which is mostly produced by *Aspergillus flavus*, *A. bombycis*, *A. nomius*, *A. parasiticus*, and *A. tamarii* in chili [23–25].

2.4 Damping off (*Pythium* spp.)

Damping off is a very common disease in nursery seedling beds showing irregular patches. *Pythium* spp. cause this disease which is basically a soil-borne fungus. This disease is more severe in soils with poor drainage.

2.5 Powdery mildew (*Leveillula taurica*)

Powdery mildew is caused by *Leveillula taurica* (asexual stage of fungus). It usually occurs in warm climates (65–95°F), both dry and humid weather. It affects the lower leaf surface, where the lesions are covered with a white to gray powdery growth. Disease progresses from the older leaves to younger leaves, and defoliation is a prominent symptom which reduces the size and number of fruits.

2.6 *Fusarium* wilt (*Fusarium oxysporum*)

Fusarium sp. is a soil-borne fungus and can survive for several years. It is usually grown in wet soil and high temperature. It causes wilting of the plant and upward and inward rolling of the leaves, and as a result leaves turn yellow and die. It causes discoloration of the vascular system of plant, particularly in the lower stem and roots.

2.7 *Cercospora* leaf spot (frog eye) (*Cercospora* spp.)

The leaf lesions due to this disease are circular about 1 cm in diameter with brown border and light gray centers. Lesions are also appearing on stem, petioles, and peduncles. It is a seed-borne fungus. It is also spread by water, wind, and leaf to leaf contact. Severely infected leaves turn yellow and drop from the plant. The fungus usually survives from one season to another on crop debris.

2.8 *Phytophthora* blight (*Phytophthora* spp.)

This disease is also known as *Phytophthora* root rot. Fungus affected all parts of plant. It commonly creates problem when soils are excessively wet either from over irrigation, heavy rain, or both. The most common symptom is a stem or collar rot by sudden wilting without foliar yellowing.

3. Common recommended cultural control

Integration of chemicals and biotic agents along with resistance for managing plant disease has been considered as a novel approach [26]. The studies of chemical and biological management are valuable to comprehend and understand biodiversity. In general, isolating, identifying, and determining structures of new metabolites are fundamental to reveal their chemical potential, a first step to use, conserve, and protect them [27]. Several attempts to control root-rot fungi could be amended including disinfestations of the soil and planting material with fungicidal chemicals, crop rotation with nonhosts of the fungus, and use of resistant cultivars [28–30].

- Crop rotation.
- Discard the weeds from the crop.
- Disease-resistant variety should be used.
- Same family crops should not be planted in the same field more than once every 3 years.
- Excessive irrigation prior to seedling emergence should be avoided and, after establishment, water should not stand in the field for more than 12 hours.
- Before sowing, the seed should be dried by artificial (machine) method or sunlight method.

4. Chemical control

- Disinfect containers, seed trays, and propagators thoroughly. Dettol is good but tea-tree oil and camomile tea make good alternatives and are safe to use on plants.
- The strobilurin fungicides azoxystrobin (Quadris), trifloxystrobin (Flint), and pyraclostrobin (Cabrio) have been labeled for the control of anthracnose fungus in chili crop.
- Spray twice at 10–15 days interval with 2.5 g mancozeb or 1 g carbendazim/lit of water for the control of *Cercospora* leaf spot.
- Soil sterilization by drenching the soil 4" deep with formaldehyde diluted 50 times with water or with some other effective chemical soil sterilant.
- Soil drenching with 1% Bordeaux mixture or 3 g copper oxychloride like Blue Copper per liter of water at 12 and 20 days after sowing is also useful.
- Seed treatment with 3 g captan or thiram per kg seed.

- Seed treatment with 4 g *Trichoderma viride* formulation combined with 6 g metalaxyl is highly effective.

- Foliar spraying with captafol at 0.2% followed by copper oxychloride at 0.25% and carbendazim at 0.1% during the start of anthracnose disease.

5. Conclusion

The yield of fruit and vegetable is reducing gradually every year due to the presence of different pests and pathogens which cause heavy losses. There is a need to find out different effective methods for the control of these diseases and reduce field crop losses. Due to the use of improper methods of disease management, different fungi particularly *Aspergillus* sp. grow on fruits and produce toxic compound (mycotoxins), which are harmful for human health. Due to the presence of high percentage (%) of these mycotoxins in different vegetables and fruit products, European Union countries and Japan governments have banned import of consignment products in their countries. It is need to improve disease management methods to prevent the growth of fungi during processing, marketing, and transportation and reduce the percentage (%) of mycotoxins in our products which increase our products' export, and thus country will get more foreign exchange.

Author details

Faisal Hussain* and Farzana Usman
Laboratory of Plant Pathology and Aerobiology, Department of Botany, Federal Urdu University of Art, Science and Technology, Karachi, Pakistan

*Address all correspondence to: faisalhussain@fuuast.edu.pk

References

[1] Ali SI, Qaiser M. A phyto-geographical analysis of the phanerogames of Pakistan and Kashmir. Proceedings of the Royal Society of Edinburgh. 1986;**98**(B):89-101

[2] Nasir JY, Rafique RA. Wild Flowers of Pakistan. Karachi: Oxford University Press; 1995. pp. 24-33

[3] Mahal MF. Study of fruit rot disease of chili caused by *Alternaria tenuis* and its control in Bangladesh dissertation. Rajshahi: University of Rajshahi; 2004. 360 p

[4] Swart AE, Holz G. *Alternaria alternata* rot of cold-stored table grapes in the Cape Province of South Africa. Phytophylactica. 1991;**23**:217-222

[5] Snowden AI. Post-Harvest Diseases and Disorders of Fruits and Vegetables. Vol. 2. Boca Raton, Florida, USA: CRC Press; 1992. p. 237

[6] Halfon-Meiri A, Rylski I. Internal mold caused in sweet pepper by *Alternaria alternata*: fungal ingress. Phytopathology. 1983;**73**:67-70

[7] McColloch LP, Worthington JT. Low temperature as a factor in the susceptibility of mature-green tomatoes to *Alternaria* rot. Phytopathology. 1952;**42**:425-427

[8] Barakai-Golan R, Kopeliovitch E. Effect of peel injury and enzymatic activity of the fruit on the tolerance of tomato genotypes of *Alternaria* infection. Acta Horticulturae. 1989;**258**:631-637

[9] Bruton BD, Chandler LD, Miller ME. Relationships between pepper weevil and internal mold of sweet pepper. Plant Disease. 1989;**73**:170-173

[10] Khaleeque MI, Khan SM. Fungi associated with fruit rot and die back diseases of chillies in Faisalabad. Pakistan Journal of Phytopathology. 1991;**3**:50-52

[11] Sultana N, Khanzada AK, Aslam M. A new cause of fruit rot of chillies in Pakistan. Pakistan Journal of Scientific and Industrial Research. 1992;**35**:491-462

[12] Amusa NA, Kehinde IA, Adegbite AA. Pepper (*Capsicum frutescens*) fruit anthracnose in humid forest region of South-western Nigeria. Nutrition and Food Science. 2004;**34**:130-134

[13] Tindall B. Vegetable in the Tropics. London: Macmillan Press; 1983. pp. 347-354

[14] Poulos JM. Problems and progress of chilli pepper production in the tropics. In: Hock CB, Hong LW, Rejab M, Syed AR, editors. Proceedings of the Conference on Chilli Pepper Production in the Tropics. Kuala Lumpur, Malaysia: MARDI; 1992. pp. 98-129

[15] Pakdeevaraporn P, Wasee S, Taylor PWJ, Mongkolporn O. Inheritance of resistance to anthracnose caused by *Colletotrichum capsici* in *Capsicum*. Plant Breeding. 2005;**124**:206-208

[16] Robert PD, Pernezny KL, Kucharek TA. Anthracnose caused by Colletotrichum spp., on pepper. University of Florida Extension Note; 2001. p. 104. www.edis.ifas.ufl.eduS, December 2007

[17] Aziz NH, Youssef YA, El-Fouly MZ, Moussa LA. Contamination of some common medicinal plant samples and spices by fungi and their mycotoxins. Botanical Bulletin of Academia Sinica. 1998;**39**:279-285

[18] Christensen CM, Fause HA, Nelson GH, Bates F, Mirocha CJ. Microflora of black and red pepper. Cooperative

Extension Circular. Applied Microbiology. 1967;**15**:622-628

[19] Scott PM, Kennedy BPC. Survey of ground black, white and *capsicum* peppers for aflatoxins. Journal of the Association of Official Agricultural Chemists. 1973;**56**:1452-1457

[20] Flannigan B, Hui SC. The occurrence of aflatoxin producing strains of *Aspergillus flavus* in the mold flora of spices. The Journal of Applied Bacteriology. 1976;**41**:411-418

[21] Seenappa M, Stobbs LW, Kempton AG. *Aspergillus* colonization of Indian red pepper during storage. Phytopathology. 1980;**70**:218-222

[22] Patel S, Hazel CM, Winterson AGM, Morthy E. Survey of ethnic foods for mycotoxins. Food Additives and Contaminants. 1996;**7**:833-841

[23] Kurtzman CP, Horn BW, Hesseltine CW. *Aspergillus nomius*, a new aflatoxin producing species related to *Aspergillus flavus* and *Aspergillus tamari*. Antonie Van Leeuwenhoek. 1987;**53**:147-158

[24] Goto T, Ito Y, Peterson SW, Wicklow DT. Mycotoxin production ability of *Aspergillus tamari*. Mycotoxins. 1997;**44**:17-20

[25] Peterson SW, Ito Y, Horn BW, Goto T. *Aspergillus bombycis*, a new aflatoxigenic species and genetic variation in its sibling species *A. nomius*. Mycologia. 2001;**93**:689-703

[26] Papavizas GS. Status of biological control of soil borne plant pathogens. Soil Biology and Biochemistry. 1973;**5**:709

[27] Castillo M. Now what?? In: Muñoz O, editor. Química flora of Chile. University of Chile, Department T? Technician of researching. Santiago, Chile; 1992. 341 p

[28] Jones JP, Jones JB, Miller W. Fusarium wilt on tomato. Fla. Dept. Agric. and Consumer Serv., Div. of Plant Industry. Plant Pathology Circular No; 1982. 237 p

[29] Agrios GN. Plant Pathology. 5th ed. Elsevier Academic Press; 2005. pp. 510-514

[30] Smith IM, Dunez J, Phillips DH, Lelliott RA, Archer SA. European Handbook of Plant Diseases. Oxford: Blackwell Scientific Publications; 1988. 583 p

Section 3

Abiotic Stress

Chapter 5

The Effect of Climate Change on Abiotic Plant Stress: A Review

Okoro Gideon Onyekachi, Onu Ogbonnaya Boniface, Ngasoh Felix Gemlack and Namessan Nicholas

Abstract

The increase in the carbon-dioxide (CO_2) present in the atmosphere as a result of human activities affects the ambient temperature, and rainfall pattern in terms of season, duration, intensity of sunshine, increased drought periods, waterlogging, and increased evapotranspiration. This influence negatively the development, yield and quality of the plants grown under this condition. The quests to produce stress tolerant/resistant plants and increase crop productivity have led to the study of plant stresses, their response to different stress type and stress management procedures in plants. This chapter has discussed in details the different abiotic stressors in plants and how they are being influenced by climate change, the response of these plants to different abiotic stresses or a combination of stresses, and abiotic stress management.

Keywords: abiotic plant stress, climate change, abiotic stress management

1. Introduction

The effects of climate change on the general environment have been very significant over the years, most especially in agricultural plants grown in these environments. According to Intergovernmental Panel on Climate Change (IPCC), stresses are incurred on plants as a result of changes in the climatic conditions of the environment, which has been concluded to be the most influencing factor affecting agricultural production at low lowlands mostly occupied by developing countries [1]. Climate change increases the presence of carbon dioxide in the air and also the temperature of the environment [2]. These are major constraints to food supply and balanced environment, leading most researchers into looking for good adaptation strategies for plants under this condition [3], by developing plant species and smart crops that are resilient to the effect of climate change [4]. Vulnerability to this kind of stresses inhibited on plants due to climate change result to disruption in the plant metabolism processes, thereby incurring reduction in aptness and quality of agricultural crop production [5]. The individual climate change inducing stressors are abiotic in nature [6], and they impose stress on different plant species: these abiotic plant stressors include drought, elevated CO_2, temperature (low and high) [7], waterlogging, rainfall and sunshine intensity, chemical factors (heavy metals and pH). The fight against these stress factors are however complex due to how interrelated they are [8], with the major challenge being to understand how these plants react to different stressors, the diversity response pathways elicited by them and their genetic determination [1, 9, 10]. This chapter therefore is aimed at reviewing the effects of climate change on abiotic plant stress, plant responses to the different stress factors, and management strategies.

2. Major abiotic stressors in plants

2.1 Temperature

Development and growth of plants is hinged on the environmental temperature. This varies depending on the crop grown [11]. The effect of temperature on the potentials of plants growth are maximum and minimum daily temperature and variation in the day and night temperature with their respective average temperatures [12]. A significant rise in the temperature due to climate change over time will result in reduced rainfall, wind speed and snow cover which will also lead to a reduced length of growing season for plants, and will affect negatively the quality of the crops and agricultural productivity [13]. Fluctuations in the environmental temperature occur more speedily than changes in stress factors thus exasperating the effects of other stresses on plants [8]. The rise in temperature is a result of the global warming which varies regionally. Prospectively, the developing countries are more vulnerable and this may perhaps result to increased food insecurity in such region. A recent study on the effects of frost and extreme temperatures on wheat production (*Triticum aestivum L*) showed that frost caused unfruitfulness and abortion of formed grains while excessive heat resulted in reduction in the number of grains formed in the filling period of the grain [14]. With the notable effects of these extreme weather events on crop production which results from climate change, the impending effect of climate variation will endanger the production trend of crops thereby giving room for food insecurity. Djanaguiraman et al. [16] studied the effect of high temperature stress on a pearl millet, and they identified sensitive stages of growth of the plant, with parameters like temperature thresholds, genetic variability and fertility of pollen germination being determined. It was found that exposure to high temperature stress reduces pollen germination, seed yield, and also affects the fertility of the pollen and pistil tissues [15, 16].

The response of plants to this class of stress depends on the type of crop or plant as seen in **Figure 1**. Broccoli (*Brassica oleracea var. italica*) growth response to temperature started from 5°C and retired at 25°C while that of maize (*Zea mays* L.) started from approximately 10°C and retired at approximately at 40°C.

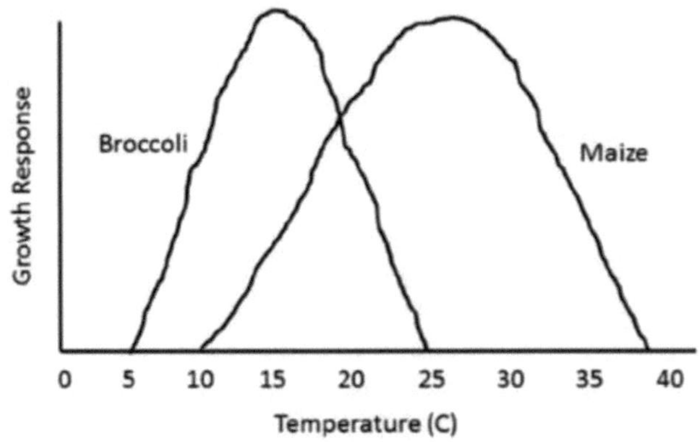

Figure 1.
Minimum and maximum temperature response for maize and broccoli [11].

2.2 Drought and rainfall pattern

Drought is one of the major abiotic stressors of agricultural plants restraining crops' return globally [17]. It does not only affect the crops' growth, but it also affects the quality of yield. In an experiment carried out on *miscanthus* for biofuel production, drought treatment was observed to have declined plant weight by 45% and the composition of the biomass and cell wall structural rigidity were severely altered due to stress incurred by the plant during the drought period [18]. According to Dhankher [15], drought as an abiotic stress factor is being projected to hinder productivity in more than 50% of the arable lands in the world by the next 50 years. But the availability of saline water might lighten the world's water problem if plants that are salt tolerant are being developed [15, 19–22]. However, in response to this kind of plant stress, the need for drought tolerant plant is a necessity to boycott or reduce its negative impacts on food security. **Figure 2** shows an extreme case of drought on a land.

Also according to [23], there is a noticeable variability in soil water content in tropical countries because of the rainfall pattern and distribution in these areas. This is an indication that the soil water content is becoming scarce for plants use. Moreover, in an investigation to know the effect of timing precipitation on grassland and rainforest in USA, the result showed that soil water content available for plant use depends on the amount of precipitation in the study area [24]. That is, when rainfall distribution uniformity is low, the available soil water content would reduce, hence will not uniformly meet the plant's soil-water-nutrient need, thus inducing stress on plants in those affected areas. This often results due to climate change. According to [25, 26], climate change alternate the timing of rainfall from one season or period to the other either resulting in smaller precipitation event or large one depending on the shift. This sudden shift in the precipitation affect plant growth and causes plant stress by disturbing plant metabolism, arresting photosynthesis, and may finally cause plants to die off.

Change in the soil water content and soil features has a notable impact on the plant and soil processes as indicated in **Figure 3**. The response in above-ground net primary productivity (ANPP) depends on which season receives extra water and which one receives less. The findings here further showed that the soil water content determines the response of above-ground net primary productivity [27]. Generally effects of drought stress on plants are;

- Reduction in seed germination and development
- Poor growth in vegetation
- Poor reproductive growth
- Reduction in leaf weight
- Reduced photosynthesis
- Reduced stomatal conductance and
- A significant reduction in the total dry matter.

2.3 Waterlogging/flooding

Climate change has altered the hydrological cycle processes which have resulted to impairment or reduction in crop growth in so many areas around the world.

Figure 2.
Xerohalophytes growing in soil impacted by severe drought [15].

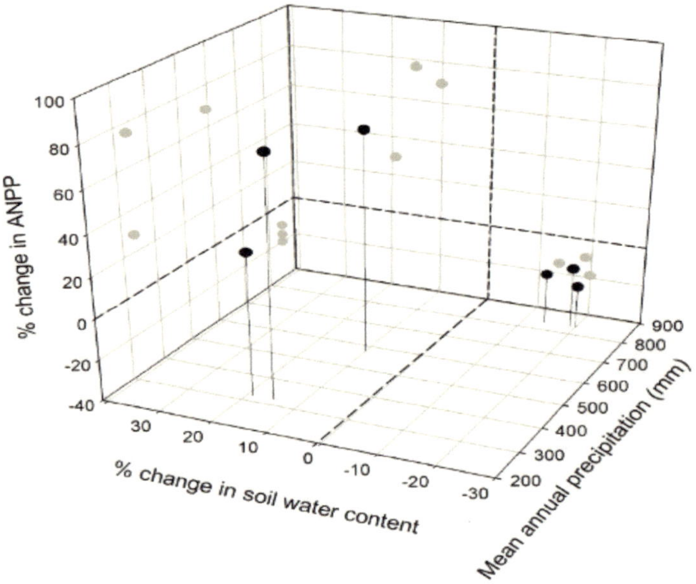

Figure 3.
The effect of precipitation at extreme level on the above-ground net primary productivity through a rainfall slope [18].

Consequently, there is a large scale reduction in agricultural production especially on a flatland or places near the river due to waterlogging [28]. Waterlogging occur as result of leakage from irrigation canals or pure surface drainage but predominantly caused by heavy rainfall in such area [29]. As a result, there is increased soil compaction, also the available oxygen (O_2) for plants cells are reduced because the diffusion process of O_2 is slow in ponding water [29, 30]. Due to the limited supply in O_2, anaerobic bacteria releases venomous amount of iron ion, manganese ion and sulfide [29].

Crops grown in a waterlogged condition undergo different physiological and morphological variations [29]. One of the major response of plant to waterlogging stress is stomatal cessation which impacts not only gas exchange, but also reduces the submissive absorption of H_2O, which is harmfully prejudiced by anaerobic conditions in the rhizosphere [28]. Transpiration is also reduced which eventually results to wilting of the leaf and early senescence; consequentially, foliar abscission will become the result [31]. In flood logged area, respiration of roots are not aerated in which gas diffusion are severely reserved, thereby resulting to it changing from aerobic to anaerobic conditions which is dangerous to plants development [31, 32]. Waterlogging also deters the nutrients intake of plants as presented in **Table 1** [33].

Nutrients	Wheat			Barley		
	Control	Waterlogged	SC	Control	Waterlogged	SC
Mg/g dry matter						
N	47.1	38.1	30–45	49.9	34.8	20–40
P	6.2	4.9	3.5–6.0	5.1	3.9	3.5–6.0
K	57.4	48.6	33–45	63.2	45.1	30–55
Ca	6.3	5.8	4–10	8.3	6.9	5–10
Mg	1.9	1.4	2–3.5	2.3	1.9	1.5–3.0
Mg/kg dry matter						
Mn	41.8	27.5	35–100	37.9	21.9	30–100
Cu	12.2	10.0	7–15	10.5	7.2	6–12
Fe	92.8	89.7	–	89.9	69.1	–
Zn	39.6	28.5	25–70	38.4	26.5	20–60
Mg/g dry matter						
N	49.1	29.0	30–45	48.9	34.5	20–40
P	4.6	3.3	3.5–6	4.4	3.6	3.5–6.0
K	50.5	28.0	33–45	54.5	28.1	30–55
Ca	4.9	6.2	4–10	8.1	12.2	5–10
Mg	3.2	2.0	2–3.5	3.0	2.8	1.5–3.0
Mg/kg dry matter						
Mn	53.9	35.1	35–100	63.9	47.7	30–100
Cu	10.7	7.9	7–15	10.1	9.4	6–12
Fe	72.1	70.0	–	69.5	71.0	–
Zn	38.6	21.4	25–70	40.5	34.1	20–60

Table 1.
Mineral concentrations of spring barley and spring wheat shoots as related to a 15 days period of waterlogging.

An investigation to know the effect of waterlogging on growth nutrient concentrations observed that there was high sprout dry weight of wheat in the waterlogged treatments compared to the control. Both in barley and wheat there was an indication of reduction in nutrients such as Nitrogen, Phosphorus, Potassium, Manganese, Copper, Zinc and Magnesium as shown in **Table 1**.

2.4 Salinity

The effect of salinity on agricultural crop production and food supply according to [15] has been on the increase worldwide, with the cultivation of salt-sensitive crops such as rice and wheat being a worldwide practice [34], salinity stress needs to be promptly addressed. Rice according to [34] is one of the crops planted most around the world as it is a major source of food for almost all humanity. Thirty-nine (39) million hectares out of the 130 million hectares of land on which rice is cultivated worldwide contains a certain degree of salinity which varies with the rice cultivar [35, 36]. Salinity is a stressor common to arid and semi-arid regions of the world where evapotranspiration exceeds rainfall, and as a result leads to inadequate rain to filter away the soluble salts from the root zone [37]. Lands with salinity stress problem covered at least 7% of the world land area four decades ago

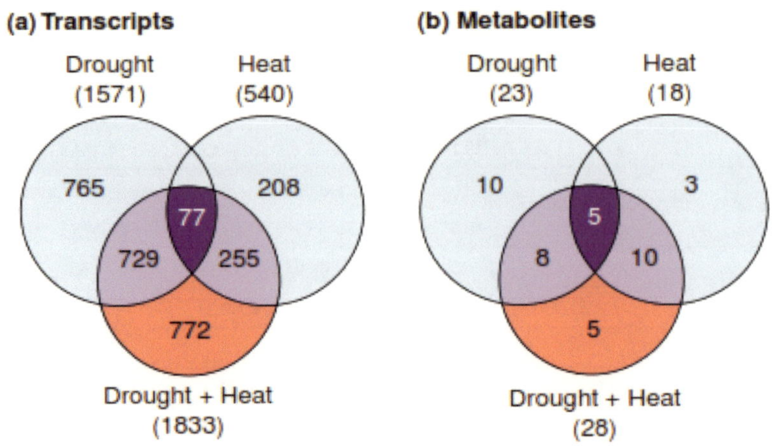

Figure 4.
Unique molecular characteristics of drought and heat stress combination [6].

according to [37], and has been seen to double every two decades. Showing that the salinity stress problem is rapidly increasing every day, majorly as a result of climate change [38, 39]. Approximately 1.6 mha of land is being lost to salinity stress every year, with 60% salinization coming from natural sources (weathering of minerals, and soils developed from saline rocks) and 40% coming from secondary sources (irrigation, deforestation, overgrazing or intensive cropping). But ironically majority of the land lost as a result of secondary salinization is caused as an adverse effect of irrigation of the farmland [37] with both normal water and treated sewage effluents. This takes place in at least 20% of irrigated lands in agriculture [38]. The salinity stressor reduces drastically the ability of plants to take up water and other nutrients from the soil, leading to stunted growth; salt deposits find their way into the transpiration stream and damage the cells in leaves thereby causing leaf burn; it also affects the enzyme activity happening within the plant. The salinity of a soil is measured in terms of electrical conductivity (EC), and these soils have a mixture of salts of Sulfate, Sodium, Magnesium, Chlorine and Calcium.

But most often, it is the combination of two more stressor that affect growth of a plant, for example heat and salinity or drought and heat as indicated by the Venn diagram presented in **Figure 4** [6]. Several researches have investigated the combine effects of drought and heat stress on maize, sorghum and turf-grasses growth and productivity. The studies showed that the combine effects of the two stressors were higher compared to when the stresses were applied individually [40–42].

3. Plant responses to abiotic stress

Having seen that abiotic stresses on plants like temperature, flooding, drought, and salinity has a serious effect on crop production, as they cause stunted growth in plants, affect plant metabolism and thus reduce crop yield by as much as 70% [43], it has also been seen that 50% of crop yield losses come as a result of abiotic stress in most crop plants [43]. Also with the constant changes in the climatic conditions of the world, most of these abiotic stresses on plants are made worse than how it was in previous years. For example, the increased presence of CO_2, methane, nitrous oxide and chlorofluorocarbons in the atmosphere as a result of industrialization and other human activities have increased drastically the average ambient temperature which causes heat stress [43, 44]. This increase in the ambient temperature of the environment also increases evaporation in water bodies, thereby causing an unexpected decrease in the available water for crop production which in extreme cases leads to drought [4, 34, 43]. Therefore a better understanding of plant responses to these abiotic stresses and their tolerance mechanism will help improve crop resistance and productivity [45, 46].

Plants response to abiotic stresses depends heavily on their developmental stage. Their response to drought and flooding according to [43] is majorly ordered by genetic expression and biochemical metabolism via different physiological processes. Some of the morphological growth indices affected by drought stress is the leaf area of the plant, the plant height, dry matter and biomass production. The photosynthetic efficiency, gas exchange measurements, the stomatal conductance, the phytochemical and the non-phytochemical quenching, malondialdehyde (MDA) and relative water content are the physiological indices put in check during the evaluation of drought stress in plants [47]. Therefore plants undergoing this type of stress gradually losses water from the protoplasm, the plant cell will dehydrates and this affects the structure of the plant cell [43, 45, 48]. When all the free water in the protoplasm of the plant cell is lost, only plants that are desiccation-tolerant (resurrection plants) will remain, and others will die off [43].

Plant's architecture is altered during flooding, although this is majorly dependent on the type of flood. When the plants are covered completely with floodwater, plant growth is stopped to conserve energy to be used once the level of water goes down. This strategy is important when the plant is completely flooded [43, 45]. In situations whereby the plants are completely flooded for a long period of time, the plant dies off as a result of the food reserve being completely depleted. During this period, there is also low oxygen (O_2), and the plants resort to rearranging their anatomy and metabolism so they can be able to function. This rearrangement to create space and a continuous gas channel to facilitate an internal O_2 from the canopy to the root of the plant is known as aerenchyma [45]. Adjustment to the metabolic pattern of the plants during flood stress is driven by hormones like the abscisic acid, gibberellin, and ethylene.

In periods of high temperature, there is a very severe damage to the chlorophyll, as heat stress changes the structural arrangement of the thylakoids, thus affecting its functionality, and also reducing the chlorophyll content of the plant. The above adverse effects on the plant reduces the ability of the plant to photosynthesize because by reducing the chlorophyll content, the photosynthetic pigments are reduced, thereby leading to physiological impairment and reduced growth of the plant [49]. The leaf of the plant also loses water under high temperatures thus leading to stomatal closure due to reduced leaf water potential [43, 49]. According to [50] stomatal closure is the major factor affecting photosynthesis in plants. High temperatures according to [51] cause impaired pollen and ovary development which adversely leads to the bad reproductive health of plants. Plants also undergo denaturation of protein and enzymes under extreme temperature conditions which gives rise to the programmed death of the tissues and cells [43, 52].

4. Abiotic stress management

Managing abiotic stresses in plants is a very is a challenging task because of its multifaceted nature. Comprehensive researches for the development of abiotic stress tolerance in plants are in progress, comprising genes from several pathways like the osmolyte synthesis, ion homeostasis, antioxidative pathways, and regulatory genes [53]. Several attempts have been made to substantiate the role of "single-function" gene(s) in the past ten (10) years, as well as transcription factor(s) for abiotic stress tolerance [53]. Because abiotic stress tolerance is multigenic in nature, the modern trend is moving towards genetic transformation of multiple genes or transcription factors. Generally, abiotic stresses can be managed by genetically improving the genes and transcription factors, or by employing cultural practices which includes modification and adjustment of planting time and crop density in the field so as to contain the effect of these abiotic stresses. Another method is to apply phytohormones, signaling and trace elements and also by applying osmoprotectants [44, 51].

Conventionally, some of the plants tolerant to these abiotic stresses discussed in the previous sections can be recognized when proper screening methods and criteria for selection are employed via cultural field breeding techniques, like modifying planting time and adjusting the compactness of plants cultivated to evade these stress situations [43, 52]. One major limitation in adopting the conventional field screening of these plants is that the screening and identification process of a plant for a particular type of stress-tolerant trait might be severely influenced by other stress factors. For instance, the quality of the screening process of a plant for cultivars tolerant to extreme temperature stresses, might be affected by other stresses either biotic (pests and pathogens) or abiotic (salinity). Therefore [44, 52] was of the opinion that glasshouses screening should be employed as an alternative to field screening [43]. The selection of some stress-tolerant plants for cultivation has notably been hindered by the polygenic nature of the trait, epistasis, and genotype by environment and low heritability [43]. These traits can be identified by carrying out quality trait loci (QTL) analysis. The qualitative trait loci (QTL) analysis according to [43, 44, 52] were carried out on rice, maize and wheat and heat stress tolerant traits were found, although these traits were found to be associated with the reproductive stage of these plants [43, 54, 55]. Some other loci, like the loci for freezing tolerance at vegetative stage, chilling at seedling stage, submergence-tolerance loci was also found in Arabidopsis model plants, maize and rice.

Biotechnological advancement has recently given a way to genetically modify the crop plants to be resistant or tolerant to some stress factors [56]. Advancement in biotechnology and genetic engineering brought out tools and methods of controlling the mechanism of abiotic stress resistance, and for developing crops tolerant to specific stress by introgression of the genes involved in stress response and putative tolerance [57]. This is important as it introduces the modified genes into genomes of elite breeding lines, and as such helps to improve crop varieties [16, 52, 54, 56, 57].

5. Conclusion

It can be seen that climate change severely intensifies the effect of these abiotic stresses on plants. The stresses not properly managed make it impossible to meet global food demand. The tolerance of plants to abiotic stress is as a result of activities of diverse stress-responsive genes that relate with other elements of stress-transducing pathways. A very good knowledge of the physiological processes in response to different abiotic stress will help design methods and mechanisms to transform in order to improve tolerance of different abiotic stress in plants. Thus, stress-tolerant gene resources from crop plants, model plants and microorganisms need to be identified and developed by making use of appropriate genetic transformation technologies.

Author details

Okoro Gideon Onyekachi[1]*, Onu Ogbonnaya Boniface[1], Ngasoh Felix Gemlack[2] and Namessan Nicholas[2]

1 Department of Agricultural and Bioresources Engineering, University of Nigeria, Nsukka, Enugu State, Nigeria

2 Department of Agricultural and Bioresources Engineering, Taraba State University, Taraba State, Nigeria

*Address all correspondence to: gideononyekachi1990@gmail.com

References

[1] Andy P. Abiotic stress tolerance in plants. Plant Science. 2016;7:1-9

[2] Hirayama T, Shinozaki K. Research on plant abiotic stress responses in the post-genome era: Past, present and future. The Plant Journal. 2010;**61**:1041-1052

[3] Rosenzweig C, Elliott J, Deryng D, Ruane AC, Müller C, Arneth A. Assessing agricultural risks of climate change in the 21st century in a global gridded crop model intercomparison. Proceedings of the National Academy of Sciences of the United States of America. 2014:3268-3273

[4] Wheeler T, Von Braun J. Climate change impacts on global food security. Science Direct. 2013;**341**:508-513

[5] Rejeb IB, Pastor V, Mauch-Mani B. Plant responses to simultaneous biotic and abiotic stress: Molecular mechanisms. Plant, Cell & Environment [Review]. 2014;**3**:458-475

[6] Mittler R. Abiotic stress, the field environment and stress combination. Trends in Plant Science. 2006;**11**(1):15-19

[7] Compant S, Van Der Heijden MG, Sessitsch A. Climate change effects on beneficial plant–microorganism interactions. FEMS Microbiology Ecology. 2010;**73**(2):197-214

[8] Ashraf M, Foolad MR. Roles of glycine betaine and proline in improving plant abiotic stress resistance. Environmental and Experimental Botany. 2005;**59**:206-216

[9] Bellard C, Bertelsmeier C, Leadley P, Thuiller W, Courchamp F. Impacts of climate change on the future of biodiversity. Ecological Letters. 2012;**15**

[10] Wallace JS, Acreman MC, Sullivan CA. The sharing of water between society and ecosystems: From conflict to catchment–based co–management. Philosophical Transactions of the Royal Society of London. Series B, Biological Sciences. 2003:2011-2026

[11] Hatfield JL, Prueger JH. Temperature extremes: Effect on plant growth and development. Weather and Climate Extremes. 2015;**10**:4-10 (13/10/2018)

[12] Essay U. Effect of Temperature on Plant Growth. 2013. Available from: https://www.ukessays.com/essays/environmental-sciences/effect-temperatures-plant-growth-8273.php?vref=1

[13] Naglaa AA, Moses V, Prakash CS. Climate change effects on abiotic stress in plants. GM Crop Food. 2014;**5**:77-80

[14] Barlow K, Christy B, O'leary G, Riffkin P, Nuttall J. Simulating the impact of extreme heat and frost events on wheat crop production: A review. Field Crops Research. 2015;**171**:109-119

[15] Dhankher A. Climate Resilient Crops for Improving Global Food Security and Safety. Vol. 2018. John Wiley and Sons Ltd; 2018

[16] Djanaguiraman M, Perumal R, Ciampitti IA, Gupta SK, Prasad PVV. Quantifying pearl millet response to high temperature stress: Thresholds, sensitive stages, genetic variability and relative sensitivity of pollen and pistil. Plant, Cell and Environment. 2017;**41**:993-1007

[17] Tuberosa R, Grillo S, Ellis RP. Unravelling the genetic basis of drought tolerance in crops. In: Toppi LS, Pawlik-Skowronska B, editors. Abiotic Plant Stresses. Dordrecht: Springer; 2003

[18] Van-Der-Weijde T, Huxley LM, Hawkins S, Sembiring EH, Farrar K,

Dolstra O, et al. Impact of drought stress on growth and quality of miscanthus for biofuel production. GCB Bioenergy. 2017;**9**:770-782

[19] Kerr TC, Abdel-Mageed H, Aleman L, Lee J, Payton P, Cryer D, et al. Ectopic expression of two AREB/ABF orthologs increases drought tolerance in cotton (*Gossypium hirsutum*). Plant, Cell and Environment. 2017;**41**:898-907

[20] Lakra N, Kaur C, Anwar K, Pareek S, Pareek A. Proteomics of contrasting rice genotypes: Identification of potential targets for raising crops for saline environment. Plant, Cell and Environment. 2017;**41**:947-969

[21] Herzog M, Konnerup D, Pedersen O, Winkel A, Colmer T. Leaf gas films contribute to rice (*Oryza sativa*) submergence tolerance during saline floods. Plant, Cell and Environment. 2017;**41**:885-897

[22] Pérez-Jiménez M, Hernández-Munuera M, Piñero MC, López-Ortega G, del Amor F. Are commercial sweet cherry rootstocks adapted to climate change? Short-term waterlogging and carbon dioxide effects on sweet cherry. Plant, Cell and Environment. 2017;**41**:908-918

[23] Smith MD. An ecological perspective on extreme climatic events: A synthetic definition and framework to guide future research. Journal of Ecology. 2011;**99**:656-663

[24] Zeppel MJB, Wilks JV, Lewis JD. Impacts of extreme precipitation and seasonal changes in precipitation on plants. Biogeosciences. 2013;**11**:3083-3093

[25] IPCC. Summary for policymakers. In: Field CB, Barros V, Stocker TF, Qin D, Dokken D, Ebi KL, Mastrandrea MD, Mach KJ, Plattner GK, Allen SK, Tignor M, Midgley PD, editors. Intergovernmental Panel on Climate Change Special Report on Managing the Risks of Extreme Events and Disasters to Advance Climate Change Adaptation. Cambridge, UK/New York, NY, USA: Cambridge University Press; 2011

[26] Pitman AJ, Perkins SE. Regional projections of future seasonal and annual changes in rainfall and temperature over Australia based on skill-selected AR(4) models. Earth Interactions. 2008;**12**:1-15

[27] Heisler-White JL, Blair JM, Kelly EF, Harmoney K, Knapp AK. Contingent productivity responses to more extreme rainfall regimes across a grassland biome. Global Change Biology. 2009;**15**:2894-2904

[28] Aldana F, García PN, Fischer G. Effect of waterlogging stress on the growth, development and symptomatology of cape gooseberry (*Physalis peruviana* L.) plants. Rev. Acad. Colombia Ciene. 2014;**38**:393-400

[29] Ashraf M, Mehmood S. Effects of waterlogging on growth and some physiological parameters of four Brassica. Plant and Soil. 1990;**121**:203-209

[30] Taiz L, Zeiger E. Plant Physiology. 5th ed. Sunderland, MA: Sinauer Association Inc; 2010

[31] Ashraf MA. Waterlogging stress in plants: A review. African Journal of Agricultural Research. 2012;7:1976-1981

[32] Akula R, Ravishankar GA. Influence of abiotic stress signals on secondary metabolites in plants. Plant Signaling & Behavior. 2011;**6**:1720-1731

[33] Steffens D, Hütsch BW, Eschholz T, Lošák T, Schubert S. Water logging may inhibit plant growth primarily by nutrient deficiency rather than nutrient toxicity. Plant Soil and Environment. 2005;**15**:545-552

[34] Food and Agricultural Organisation of the United Nations. The Future of Food and Agricultural Trends and Challenges. Rome: Food and Agricultural Organisation; 2015

[35] Wang W, Vinocur B, Altman A. Plant response to drought, salinity, and extreme temperatures: Towards genetic engineering for stress tolerance. Planta. 2003;**218**:1-14

[36] Vinocur B, Altman A. Recent advances in engineering plant tolerance to abiotic stress: Achievements and limitations. Current Opinion in Plant Sciences. 2005;**16**:1-10

[37] Vijayalakshmi D. Abiotic stresses and its management In agriculture. In: Stresses in Plants. India: Springer; 1999

[38] Flowers TJ, Yeo AR. Breeding for salinity resistance in crop plants—Where next. Australian Journal of Plant Physiology. 1995;**22**:875-884

[39] Massoud F. Salinity and Alkalinity as Soil Degradation Hazards. Rome: Food and Agricultural Organisation; 1998

[40] Craufurd PQ, Peacock JM. Effects of heat and drought stress in Sorghum. Experimental Agriculture. 1993;**29**:77-86

[41] Jiang Y, Huang B. Drought and heat stress injury to two cool season turfgrasses in relation to antioxidant metabolism and lipid peroxidation. Crop Science. 2001;**41**:436-442

[42] Wang ZL, Huang BR. Physiological recovery of Kentucky bluegrass from simultaneous drought and heat stress. Crop Science. 2004;**44**:1729-1736

[43] Arun-Chinnappa KS, Ranawake L, Seneweera S. Impacts and management of temperature and water stress in crop plants. In: Minhas PS, editor. Abiotic Stress Management for Resilient Agriculture. Australia: Springer; 2017. pp. 221-233

[44] Wahid A, Gelani S, Ashraf M, Foolad MR. Heat tolerance in plants; an overview. Environmental and Experimental Botany. 2007;**61**:199-223

[45] Bray EA, Bailey-Serres J, Voesenek LACJ. Response of Plants to Abiotic Stresses. Rockville: American Society of Plant Physiologists; 2000

[46] Fernando N, Mnailil S, Florentine S, Chauhan B, Seneweera S. Glyphosate resistance of C3 and C4 weeds under rising atmospheric CO_2. Frontiers in Plant Science. 2016;**7**:910

[47] Jenks MA, Hasegawa PM. Plant Abiotic Stress. 2005;**1**

[48] Rejeb IB, Pastor V, Mauch-Mani. Plant responses to simultaneous biotic and abiotic stress: Molecular mechanisms. Plants. 2014;**2014**:458-475

[49] Xu Z, Zhou G, Shimizu H. Plant responses to drought and rewatering. Plant Signaling and Behaviour. 2010;**5**:649-654

[50] Greer DH, Weedon MM. Modelling photosynthesis response to temperature of grape vine (*Vitis vinifera* cv. Semillon) leaves on vines grown in a hot climate. Plant Cell and Environment. 2012;**35**:1050-1064

[51] Hassanuzzaman M, Hossain MA, Fujita M. Physiological and biochemical mechanisms of nitric oxide induced abiotic stress tolerance in plants. American Journal of Plant Physiology. 2010;**5**:294-324

[52] Hassanuzzaman M, Nahar K, Fujita M. Extreme Temperature Responses, Oxidative Stress and Antioxidant Defense in Plants. IntechOpen; 2013

[53] Agarwal PK, Shukla PS, Gupta SK, Jha B. Genetic engineering of plant genes. Molecular Biotechnology. 2013;**54**:103-123

[54] Collins NC, Tardieu F, Tuberosa R. Quantitative trait loci and crop performance under abiotic stress; where do we stand? Plant Physiology. 2008;**147**:469-486

[55] Talukder S, Babar M, Vijayalakshmi K, Poland J, Prasad PVV, Bowden R. Mapping qualitative trait loci for the traits associated with heat tolerance in wheat. BMC Genetics. 2014;**15**:97

[56] Hassanuzzaman M, Nahar K, Alam MGM, Fujita M, Roychowdhury R. Physiological, biochemical and molecular mechanisms of heat stress tolerance in plants. International Journal of Molecular Science. 2013;**14**:9643-9684

[57] Bhatnaga-Mathur P, Vadez V, Sharma KK. Genetic modifications for stress tolerants in plants. Plant Cell Repository. 2007;**27**:411-424

Chapter 6

Nanotechnology a Potential Tool to Mitigate Abiotic Stress in Crop Plants

Aparajita Das and Bimal Das

Abstract

The response of plants to abiotic stress is complex and involves changes in their morphology, physiology and metabolism. A number of strategies are being followed to enhance the tolerance of abiotic stress conditions, including the development of genetically-engineered varieties containing various gene constructs believed to enhance the performance under stress conditions. Nanotechnology is a versatile field and has found application in almost all the existing fields of science. The application of nanoparticles increased germination and seedling growth, physiological activities including photosynthesis and nitrogen metabolism, leaf activities of CAT, POX and APX, chlorophyll contents, protein, carbohydrate contents and yield, and also positive changes in gene expression indicating their potential use in crop improvement. Nanoparticles enhances the water stress tolerance via enhancing root hydraulic conductance and water uptake in plants and showing differential abundance of proteins involved in oxidation-reduction, ROS detoxification, stress signaling, and hormonal pathways. The mobility of the nanoparticles is very high, which leads to rapid transport of the nutrient to all parts of the plant. In particular, the most actual is to find ways to increase the adaptation potential of cultivated plants with the use of nanopreparations in stressful conditions.

Keywords: nanoparticles, abiotic stress, drought, salinity, ROS, crop plants

1. Introduction

World population is increasing day by day and by 2050 it is expected to reach 9.1 billion, but agricultural production is not rising at a parallel pace. Raising productivity is a challenge as the area under cultivation is likely to remain constant or even decrease due to increasing pressure on land for nonagricultural uses. While increased investments and technological breakthrough can improve availability, these may not necessarily translate into increased accessibility and absorption of food. With climate change on the trail, abiotic stresses are considered to be a major constraint for sustaining crop productivity. As per one of the estimates approximately 70% of yield reduction of crops is directly or indirectly influenced by abiotic stresses [1]. Abiotic stress leads to a series of morphological, physiological, biochemical and molecular changes that adversely affect plant growth and productivity. Drought, salinity and extreme temperatures are the most prevalent abiotic stresses, threatening the global food security. Development of stress tolerant plants can be a worthwhile strategy to win over the problem of decreasing global food production. Conventional breeding methods have met with limited success in improving the stress tolerance of crop plants involving inter-specific or inter-generic hybridization. The conventional breeding approaches are limited by the complexity of stress tolerance traits, as well as the low genetic variability of yield components under stress condition and lack of efficient selection criteria. It is important, therefore, to look for alternative strategies to develop stress tolerant crops. All traditional breeding methods including selection, hybridization, polyploidy and mutation have utilized for genetic improvement of crop plants. Albeit supplementary success in history of crop improvement in agricultural crops, their yield at present reached a plateau and there exists food insecurity and poverty in many developing countries. For this purpose exploration of novel strategies and their exploitation in complement to existing traditional and advanced breeding tools is the need of the hour. Now-a-days the global demand is to increase food production with limited available resources and minimum but efficient use of fertilizer and pesticides that can check pollution in the environment which ushered in new agricultural technologies to reshape modern agriculture. Among the latest technology, nanotechnology is most promising one in the era of agriculture and plant biotechnology [2]. The application of nanoparticles or nanodevices affect various developmental stages both positive and negative impact on plant growth. Nanotechnology comprises novel properties of nanomaterial that make easy for agricultural research in crop improvement program as well as alleviation to stresses [3]. Nanotechnology has been provisionally defined as relating to materials, systems and processes which operate at a scale of 100 nanometers (nm) or less. 'Nano' usually refers to a size scale between 1 and 100 nm. Nano materials are composed of components with very small size, and these components have impacts on the properties of materials at the macro level. Nanomaterials have a relatively larger surface area when compared to the same mass of material produced in a larger form. Nano particles can make materials more chemically reactive and affect their strength or electrical properties. The particles have high surface to volume ratio that increases their reactivity and possible biochemical activity [4].

2. Types of nanomaterials used

Nanomaterials have applications in the field of nanotechnology, and displays different physical chemical characteristics from normal chemicals (i.e. silver nano, carbon nanotube, fullerene, photocatalyst, carbon nano, silica). Common types of nanomaterials include nanotubes (Single walled carbon nanotube, multi walled carbon nanotube), dendrimers, quantum dots, fullerenes, metal (Ag, Si, Au, etc.) and metal oxide (TiO_2, SiO_2, ZnO, CuO, etc.) based. Nano-scale materials can occur in nature; both natural and manmade processes (such as volcanic activity or diesel combustion) can also give rise to nanometer-sized materials unintentionally. There are two processes for nanomaterial creation including "bottom-up" processes (such as self-assembly) that create nanoscale materials from atoms and molecules, and "top-down" processes (such as milling) that create nanoscale materials from their macro-scale counterparts. Nanomaterials can be nanoscale in one dimension (e.g. surface films), two dimensions (e.g. strands or fibers), or three dimensions (e.g. particles). They can exist in single, fused, aggregated or agglomerated forms with spherical, tubular, and irregular shapes.

3. Uptake, translocation, and accumulation of nanoparticles (NPs) into the plants

The uptake of CB and MB NMs by plants is a very recent field of study. Among CB NMs, the most studied materials are the fullerene C_{70}, the fullerol ($C_{60}(OH)_{20}$) and CNTs; while the most studied MB NMs are TiO_2, Au, Ag, Cu, CeO_2, FeO, and ZnO NPs. Uptake, translocation, and accumulation of NPs depend on the species of plant and the size, type, chemical composition, functionalization, and stability of the NPs. Usually, NPs enter the plant root system through the lateral root junctions and reach the xylem through the cortex and the pericycle [5]. The mechanism behind interaction of nanoparticles with plant system is primarily based on chemical processes which create reactive oxygen species, ion cell membrane transport activity, oxidative damage and lipid peroxidation. Once enter in the plant cells NPs react with sulfhydryl, carboxyl groups and ultimately alter the protein activity. The NPs may form complexes with membrane transporters or root exudates and subsequently be transported into the plants [6, 7]. Nanomaterials move from leaves to roots, stem, and developing grain, and from one root to another. One of the main passages of uptake and transportations to the shoot and leaves of plant is the Xylem [8, 9]. The nanomaterials are capable of penetrating through the leaf cuticle and into the cell cytoplasm [10]. In the cytoplasm, the NMs may bind with different cytoplasmic organelles and interfere with the metabolic processes at that site [11]. One of the pathways also reported particle size of 20 nm Ag nanoparticles may be transported inside the cells through plasmodesmata [12, 13]. A study on generational transmission of C_{70}-NOM in rice plants and find the presence of black aggregates of C_{70} in the leaves of the second generation of the plants treated with fullerenes only in their first generation [14].

4. Plant response to abiotic stress

Among the abiotic stresses, drought, salinity, alkalinity, submergence and mineral toxicity/deficiencies are considered as major factors that contribute to decrease crop growth and productivity [15]. Plants face various environmental stresses throughout their life cycle, therefore they develop their defense against environmental stresses at various levels by modulating molecular, biochemical and physiological pathways. In order to cope these stresses, plants adopt molecular routes by appropriate alteration of gene expressions. There are several studies which indicated that nanoparticles mediated effect on plants growth and development is concentration dependent. Nanoparticles are involved in upregulating the activities of antioxidant enzymes like, SOD, CAT and POD [16].

4.1 Effect of nanoparticles on drought stresses

Water is a vital component for plant survival and essentially required for transport of nutrients, therefore water deficiency leads to drought stress, which resulted into weakened vitality of plants [17]. Nanotechnology promises the significant effort to mitigate the drought stresses. Several recent studies (**Table 1**) have evaluated nanoparticle-mediated in different stresses [18, 19].

The application of different concentrations of silica nanoparticles improves the plant tolerance toward drought stress in Hawthorns (*Crataegus* sp.), the physiological and biochemical responses varies in hawthorn seedlings to different concentrations of silica nanoparticles at different level of drought stress from

Abiotic stresses	Nanomaterial	Concentration	Plant species	Stress responses	Refs.
Drought stress	Nano TiO_2	0.01, 0.02, and 0.03%	Wheat (*Triticum aestivum* L.)	Increasing growth, yield, gluten and starch content of wheat	[20]
	Nano TiO_2	0, 10, 100, and 500 mg L^{-1}	Flax (*Linum usitatissimum* L.)	Enhancing chlorophyll and carotenoids content, improving flax growth and yield attributes, decreasing H_2O_2 and malondialdehyde (MDA) content	[21]
	Nano TiO_2	0%, 0.01% and 0.03%.	Basil (*Ocimum basilicum* L.)	Improving the negative effects of drought stress on basil plants	[22]
	Nano Zero valent Fe		*Arabidopsis thaliana* L.	Activation of plasma membrane H^+-ATPase, stomatal opening, increasing Chl content and plant biomass, maintaining normal drought sensitivity, increasing CO_2 assimilation in thale cress plants	[23]
	Nano SiO_2	0, 10, 50 and 100 mg L^{-1}	*Crataegus* sp.	A positive significant effect on photosynthetic rate, stomatal conductance and plant biomass, non-significant effect on chlorophyll and carotenoid content	[24]
	Nano ZnO	0.5, 1 g L^{-1}	Soybean (*Glycine max* L.)	Increasing germination percentage and germination rate, decreasing in seed residual fresh and dry 8 weight of soybean	[25]
	SiO_2	0, 10, 50 and 100 mg L^{-1}	Hawthorns (*Crataegus* sp.)	SNPs increased plant biomass, xylem water potential and MDA content, especially under drought conditions, RWC and ELI were not affected by the SNP pre-treatments.	[24]
	Silicon		Sorghum (*S. bicolar*)	Increase in leaf area index (LAI), specific leaf weight (SLW), chlorophyll content (SPAD), leaf dry weight (LDW), shoot dry weight (SDW), root dry weight (RDW), total dry weight (TDW)	[26]
	TiO_2 and SiO_2	25, 50, 100 and 200 ppm) or nano-SiO_2 (400, 800, 1600 and 3200 ppm)	Cotton	Increased total phenolics, total soluble proteins, total free amino acids, proline content, total reducing power, total antioxidant capacity, catalase activity peroxidase activity and superoxide dismutase activity in comparison with control	[27]
Salinity stress	Nano SiO_2	25 mM	Tomato (*Lycopersicum esculentum* L.)	Lower levels of nano-SiO_2 enhanced seed germination potential, root length and dry weight. Higher levels suppressed seed germination characteristics	[28]
	Nano SiO_2		Basil (*Ocimum basilicum* L.)	Increasing fresh and dry weight, chlorophyll content and proline content	[29]
	Nano SiO_2		Squash (*Cucurbita pepo* L.)	Improving seed germination and growth characteristics, reduced levels of MDA, H_2O_2 and electrolyte leakage, reducing chlorophyll degradation and oxidative damage, enhancing photosynthetic parameters antioxidant enzymes	[30]

Abiotic stresses	Nanomaterial	Concentration	Plant species	Stress responses	Refs.
	Nano SiO_2		Tomato (*Solanum lycopersicum* L.)	Up-regulating the expression profile of four salt stress genes and six genes were down-regulated, suppressing the effect of salinity on seed germination rate, root length and fresh weight	[31]
	Nano ZnO and Fe_3O_4	30,60,90 mg L^{-1}	Moringa peregrina	Reduction in Na^+ and Cl^- contents, increasing N, P, K^+, Ca^{2+}, Mg^{2+}, Fe, Zn, total chlorophyll, carotenoids, proline, carbohydrates, crude protein and enzymatic and non-enzymatic antioxidants	[32]
	Nano ZnO	2 g L^{-1}.	Sunflower (*Helianthus annuus* L.)	Increasing growth, net CO_2 assimilation rate, sub-stomatal CO_2 content, chlorophyll content, Fv/Fm and Zn content and decreasing Na^+ content in leaves	[33]
	ZnO	2 g L^{-1}.	Sunflower (*Helianthus annuus* L.)	Increase growth, proline content, and some antioxidant enzyme activities	[33]
Flooding stress	Nano Ag	40, 80 or 120 ppm	*Crocus sativus*	Blocking of ethylene signaling, promotion of root growth	[34]
	Nano Al_2O_3		Soybean (*Glycine max* L.)	Regulation of energy metabolism and cell death, improved growth	[35]
	Nano Ag		Soybean (*Glycine max* L.)	Reducing generation of cytotoxic byproducts of glycolysis, increasing the abundance of stress-related proteins, enhancing seedling growth	[35]

Table 1.
Effects of nanomaterials on abiotic stresses.

moderate to severe stress [24]. It was confirmed to positive effect on photosynthesis parameters, malondialdehyde (MDA), relative water content (RWC), membrane electrolyte leakage (ELI) as well as chlorophyll, carotenoid, carbohydrate and proline contents by pre-treatment of SNPs. Perhaps involvement of silicon nano particles in maintaining critical physiological and biochemical attributes in order to induce drought tolerance in hawthorn seedlings under drought stress, but exact mechanism is yet to be understood [24]. Application of silicon on two sorghum cultivars having different drought susceptibility showed improved drought tolerance irrespective of their drought susceptibility by lowering shoot to root (S/R) ratio, which perhaps suggested the improved root growth and the maintenance of the photosynthetic rate [36]. The addition of SiO_2 to plant medium reduces the penetrability of the plasma wall of the leaf cells resulting in the loss of lipid peroxidation and also, SiO_2 protects cellular wall against heat and drought stress [37]. Proline content significantly increased when silica nanoparticles were applied under stress, in comparison with common silica fertilizer [29]. Application of nano-Si caused a significantly increase in the activities of catalase (CAT) and peroxidase (POD) in plant leaves as compared to unstressed plants of faba bean [29], tomato [30] and alfalfa [38]. Furthermore, silica nanoparticles also exhibit its effect on xylem humidity, water translocation and enhance turgor pressure, thus leaf relative water content and water use efficiency will be increased in pants.

Treatment with Rutile (TiO_2) has led to increase germination, germination indices, vigor indices, plant dry weight, chlorophyll formation, activities of ribulose bisphosphonates carboxylase and oxygenase, rate of evaluation of oxygen in the chloroplast leading to promoted photosynthesis [39, 40]. TiO_2 NPs augmented wheat plant growth and yield with its components under water deficit stress condition [20] and also regulates enzymes activity involved in nitrogen metabolism such as nitrate reductase, glutamate dehydrogenase, glutamine synthase, and glutamic-pyruvic transaminase that helps the plants to absorb nitrate. The effects of nano-TiO_2 improved germination, light absorbance, photosynthetic activity and activate Rubisco [41] also promoted antioxidant stress by decreasing the accumulation of superoxide radicals, hydrogen peroxide, malonyldialdehyde content and enhance the activities of superoxide dismutase, catalase, ascorbate peroxidase, guaiacol peroxidase in spinach [42]. Nano- TiO_2 also was observed to promote the growth of spinach through an increase in photosynthetic rate and nitrogen metabolism in spinach [43]. Nano-TiO_2 can enhance plant water and nitrogen use and stimulate some antioxidant enzyme activities, such as SOD, POD and CAT such as in canola [25]. The application of nano zinc oxide has potential to increase seed germination percentage and germination rate in soybean as compared to those were subjected to water stress. It was further suggested that nano zinc oxide application under drought stress decrease seed residual fresh and dry weight, which shows that zinc nanoparticles were effective for using of seed reservoirs to seedling growth and enhance drought tolerance [44]. A study revealed the significant effect of iron nanoparticles under drought stress in plants on traits like number of boll per branch, number of seeds per boll, the 1000 seed weight and yield at probability level of 1%. Foliar application of iron nanoparticles exhibited drought stress mitigating effects on yield components and oil percentage of Goldasht spring safflower cultivars. Application of Fe nanoparticles also enhance yield and yield components at two stages of flowering and granulation, although it was better at flowering stage than seed formation in contrast to drought stress conditions without Fe nanoparticles application [45]. Advances of silver nanoparticles (AgNPs) application of silver nanoparticles (AgNPs) could be attributed toward mitigating water stress mediating loss of plant growth and yield [46].

4.2 Effect of nanoparticles in salinity stress

Salinity is the major concern of scientific community to attain sustainable crop production, it is estimated that more than 20% of cultivated land worldwide is experiencing salinity stress and the amount is increasing day by day. Salinity stress causes the negative impact on various biochemical and physiological processes which are associated with plant growth and yield. Lowering of soil osmotic potential, creation of nutritional imbalance, enhancing specific ionic toxicity (salt stress) or one or more combination of these factors, are some of the common implications of salinity stress experienced by plants [47]. The Application of nanofertilizers could be a potential approach to address such issues of soil toxicity and other associated stress problems. It is reported that silicon nanoparticles and silicon fertilizer exhibited promising effects on physiological and morphological traits on vegetative features of basil under salinity stress. It was evident from results which indicated significant increase in growth and development indices, chlorophyll content and proline level in basil (*Ocimum basilicum*) under salinity stress, when treated with silicon nanoparticles and silicon fertilizer [29]. Application of Nano-SiO_2 particles have shown potential increase in chlorophyll content, leaf fresh weight, leaf dry weight, proline accumulation and upregulated antioxidant enzymes activity under salinity stress [28]. Application of silicon nano-particles on lentil (*Lens culinaris*

Medik.) genotypes under salinity stress revealed significant increase in seed germination and seedling growth, whereas significant reduction in germination percent and seedling growth due to the salinity stress under without treatment of nanoparticles. Adding SiO_2 nanoparticles not only enhance seed germination and early seedling growth but also increase other related traits in lentil genotypes under salinity stress. Therefore, SiO_2 nano-particles ameliorate different defense mechanisms of plants against salt toxicity [48]. Other studies in maize suggested that increase in fresh soot fresh and weight under salinity stress when applied by nano SiO2 [41]. One strategy which silica nanoparticles adopts to mitigate salinity stress in plants is to reduce Na^+ ion concentration, perhaps by reducing Na^+ ion absorption by plant tissues [49]. Since primary impact of salinity stress on plant growth is due to reduction of osmotic potential and toxicity of Na^+ ion. Pure alumina nanoparticles (13 nm) without any modifications reduced root elongation in studied plants cucumber (*Cucumis sativus*), soybean (*Glycine max*), carrot (*Daucus carota*) and cabbage (*Brassica oleracea*), thus potentially retarding the growth of plants [50].

5. Effect of nanoparticles on antioxidant and molecular aspect of plants

Nanoparticles can interact with biological systems such as plants chemically or mechanically; and these specific interactions originate mainly from their small size, large surface area, and intrinsic catalytic reactivity. There are only few studies describing nanoparticles impact on antioxidant and molecular level. The treatment of silver nanoparticles in *Brassica juncea* [51] augmented the activities of antioxidant enzymes (ascorbate peroxidase, guaiacol peroxidase and catalase) which resulted in decreased level of reactive oxygen species (ROS). When, Ag NPs of 6 nm sizes were applied at the concentration of 5 mg/l, it caused activation of antioxidant system of *Spirodela polyrhiza,* evident by induced activity of superoxide dismutase, catalase and peroxidase [52]. In addition, concentration of reactive oxygen species, glutathione and malondialdehyde was also increased significantly. Application of gold nanoparticles (GNPs) in *Brassica juncea* seedlings a considerable enhancement appears in the activities of antioxidant enzymes such as, ascorbate peroxidase, guaiacol peroxidase, catalase and glutathione reductase along with higher accumulation of H_2O_2 and proline in the GNPs treated plants [53]. H_2O_2 and proline contents are found to be increased with increasing concentration of GNPs. In particular, actions of ascorbate peroxidase (APX), guaiacol peroxidase (GPX) and glutathione reductase (GR) are increased up to 400 ppm concentration of GNPs while GR activity is maximum at 200 ppm GNPs. The exposure of CeO_2 nanoparticles in kidney bean were significance responses of antioxidant enzyme (ascorbate peroxidase, catalase and guaiacol peroxidase) activities in leaf, root and stem [54]. They observed that upon prolonged exposure to 500 mg nano CeO_2/l, the root antioxidant enzyme activities were significantly reduced; simultaneously root soluble protein was increased. Moreover, guaiacol peroxidase enzyme (GPX) activity in leaf was enhanced with nano CeO_2 exposure in order to maintain cellular homeostasis. Gene expression analyses of the model plant Arabidopsis by RT-PCR have provided new insights into the molecular mechanisms of plant responses to Ag NPs. The transcriptional response of Arabidopsis plants exposed to Ag NPs was analyzed using whole-genome cDNA expression microarrays [55] which result in upregulation of 286 genes, including the genes primarily associated with metal and oxidative stress (e.g., vacuolar cation/proton exchanger, superoxide dismutase, cytochrome P450-dependent oxidase, and peroxidase), and down regulation of 81 genes, including the genes involved in plant defense system and hormonal stimuli (e.g., auxin-regulated gene involved in organ size-ARGOS, ethylene signaling pathway, and SAR against pathogens). On the other hand, the effects of silver nanoparticles on proteomic study of rice that reveals silver nanoparticles responsive proteins were primarily associated with oxidative stress response pathway, Ca^{2+} regulation and signaling, transcription, protein degradation, cell wall synthesis, cell division, and apoptosis [56]. The effect of zinc oxide (nZnO) in *Arabidopsis thaliana* [57], fullerene soot (FS) or titanium dioxide ($nTiO_2$) nanoparticles on gene expression in roots and resulted in 660 up- and 826 down-regulated genes, 232 up- and 189 downregulated genes, and 80 up- and 74 down-regulated genes, respectively (expression difference > 2-fold). The genes induced by nZnO and FS included mainly ontology groups annotated as stress responsive, including both abiotic (oxidative, salt, water deprivation) and biotic (wounding and defense to pathogens) stimuli. Application of multi-walled carbon nanotubes markedly influenced tomato seed germination and seedling growth by up-regulating stress-related gene expression [58].

6. Conclusion and perspectives

Application of nanotechnology in agriculture, even at its global level, is at its nascent stage. Nanoscience is leading to the development of a range of inexpensive nanotech applications for enhanced plant growth, biotic and abiotic stress responses. Nanoparticles enhances the stress tolerance via enhancing root hydraulic conductance and water uptake in plants and showing differential abundance of proteins involved in oxidation–reduction, ROS detoxification, stress signaling and hormonal pathways. Nanoparticles interaction with plant cell results in modification of plant gene expression and biological pathways which ultimately affect plant growth and development. Research on nanotechnology in agriculture a vast study on fabrication, characterization, standardization, biodegradability, ecofriendly nature and also possible uptake and translocation of nanoparticles by plants is needed.

Acknowledgements

The authors are grateful to Prof. Chittaranjan Kole, former Vice-Chancellor of BCKV for guiding and providing necessary information.

Conflicts of interest

The authors declare no conflict of interest.

Author details

Aparajita Das[1] and Bimal Das[2*]

1 Uttar Banga Krishi Viswavidyalaya, Pundibari, India

2 College of Agriculture (Extended Campus), Uttar Banga Krishi Viswavidyalaya, Majhian, India

*Address all correspondence to: bimal.das987@gmail.com

References

[1] Acquaah G. Principles of Plant Genetics and Breeding. Oxford, UK: Blackwell; 2007

[2] Scrinis G, Lyons K. The emerging nano-corporate paradigm: Nanotechnology and the transformation of nature, food and agri-food systems. International Journal of Sociology of Agriculture and Food. 2007;**15**:22-44

[3] Carmen IU, Chithra P, Huang Q , Takhistov P, Liu S, Kokini JL. Nanotechnology: A new frontier in food science. Food Technology. 2003;**57**:24-29

[4] Dubchak S, Ogar A, Mietelski JW, Turnau K. Influence of silver and titanium nanoparticles on arbuscular mycorrhiza colonization and accumulation of radiocaesium in *Helianthus annuus*. Spanish Journal of Agricultural Research. 2010;**8**:103-108

[5] Dietz KJ, Herth S. Plant nanotoxicology. Trends in Plant Science. 2011;**16**(11):582-589

[6] Kurepa J, Paunesku T, Vogt S, Arora H, Rabatic BM, Lu J, et al. Uptake and distribution of ultrasmall anatase TiO_2 alizarin red S nanoconjugates in *Arabidopsis thaliana*. Nano Letters. 2009;**10**(7):2296-2302

[7] Watanabe T, Misawa S, Hiradate S, Osaki M. Root mucilage enhances aluminum accumulation in *Melastoma malabathricum*, an aluminum accumulator. Plant Signaling and Behavior. 2008;**3**:603-605

[8] Birbaum K, Brogioli R, Schellenberg M. No evidence for cerium dioxide nanoparticle translocation in maize plants. Environmental Science and Technology. 2010;**44**(22):8718-8723

[9] Pola M, Tamara LC, Andrew TH. Toxicity, uptake, and translocation of engineered nanomaterials in vascular plants. Environmental Science and Technology. 2012;**46**(17):9224-9239

[10] Sharif F, Westerhoff P, Herckes P. Sorption of trace organics and engineered nanomaterials on to wet land plant material. Environmental Sciences: Processes and Impacts. 2013;**15**(1):267-274

[11] Zhang LW, Monteiro RNA. Mechanisms of quantum dot nanoparticle cellular uptake. Toxicological Sciences. 2009;**110**(1):138-155

[12] Wuncheng W. Toxicity tests of aquatic pollutants by using common duckweed. Environmental Pollutio*n*. 1986;**11**(1):1-14

[13] Unrine JM, Colman BP, Bone AJ, Gondikas AP, Matson CW. Biotic and abiotic interactions in aquatic microcosms determine fate and toxicity of ag nanoparticles. Part 1. Aggregation and dissolution. Environmental Science and Technology. 2012;**46**(13):6915-6924

[14] Lin S, Reppert J, Hu Q , Hudson JS, Reid ML, Ratnikova TA, et al. Uptake, translocation, and transmission of carbon nanomaterials in rice plants. Small. 2009;**5**:1128-1132

[15] Boyer J. S. Plant productivity and environment. Science. 1982;**218**(4571):443-448

[16] Laware SL, Raskar S. Effect of titanium dioxide nanoparticles on hydrolytic and antioxidant enzymes during seed germination in onion. International Journal of Current Microbiology and Applied Science. 2014;**3**(7):749-760

[17] Martínez-Vilalta J, Piñol J. Drought-induced mortality and hydraulic architecture in pine populations of the

NE Iberian Peninsula. Forest Ecology and Management. 2002;**161**:247-256

[18] Barrena R, Casals E, Colon J, Font X, Sanchez A, Puntes V. Evaluation of the ecotoxicity of model nanoparticles. Chemosphere. 2009;**75**:850-857

[19] Lee CW, Mahendra S, Zodrow K, Li D, Tsai YC, Braam J, et al. Developmental phytotoxicity of metal oxide nanoparticles to Arabidopsis thaliana. Environmental Toxicology and Chemistry. 2010;**29**:669-675

[20] Jaberzadeh A, Moaveni P, Moghadam HRT, Zahedi H. Influence of bulk and nanoparticles titanium foliar application on some agronomic traits, seed gluten and starch contents of wheat subjected to water deficit stress. Notulae Botanicae Horti Agrobotanici Cluj-Napoca. 2013;**41**:201-207

[21] Aghdam MTB, Mohammadi H, Ghorbanpour M. Effects of nanoparticulate anatase titanium dioxide on physiological and biochemical performance of *Linum usitatissimum* (Linaceae) under well watered and drought stress conditions. Brazilian Journal of Botany. 2016;**39**:139-146. DOI: 10.1007/s40415-015-0227-x

[22] Kiapour H, Moaveni P, Habibi D, Sani B. Evaluation of the application of gibbrellic acid and titanium dioxide nanoparticles under drought stress on some traits of basil (*Ocimum basilicum* L.). International Journal of Agronomy and Agricultural Research. 2015;**6**(4):138-150

[23] Kim JH, Oh Y, Yoon H, Hwang I, Chang YS. Iron nanoparticle-induced activation of plasma membrane H^+-ATPase promotes stomatal opening in *Arabidopsis thaliana*. Environmental Science & Technology. 2015;**4**: 1113-1119. DOI: 10.1021/es504375t

[24] Ashkavand P, Tabari M, Zarafshar M, Tomášková I, Struve D. Effect of SiO2 nanoparticles on drought resistance in hawthorn seedlings. Leśne Prace Badawcze/Forest Research Papers Grudzień. 2015;**76**(4):350-359

[25] Mahmoodzadeh H, Nabavi M, Kashefi H. Effect of nanoscale titanium dioxide particles on the germination and growth of canola (*Brassica napus*). Journal of Ornamental and Horticultural Plants. 2013;**3**:25-32

[26] Ahmed M, Hassen F, Qadeer U, Aslam MA. Silicon application and drought tolerance mechanism of sorghum. African Journal of Agricultural Research. 2011;**6**(3):594-607

[27] Magdy AS, Hazem MMH, Alia AMN, Alshaimaa AI. Effects of TiO_2 and SiO_2 nanoparticles on cotton plant under drought stress. Research Journal of Pharmaceutical, Biological and Chemical Sciences Biochemical and Physiological. 2016;**7**(4):1540

[28] Haghighi M, Afifipour Z, Mozafarian M. The effect of N-Si on tomato seed germination under salinity levels. Journal of Biological and Environmental Sciences. 2012;**6**(16):87-90

[29] Kalteh M, Alipour ZT, Ashraf S, Aliabadi MM, Nosratabadi AF. Effect of silica nanoparticles on basil (*Ocimum basilicum*) under salinity stress. Journal of Chemical Health & Risks. 2014;**4**:49-55

[30] Siddiqui MH, Al-Whaibi MH. Role of nano-SiO_2 in germination of tomato (*Lycopersicum esculentum* seeds Mill.). Saudi Journal of Biological Sciences. 2014;**21**:13-17

[31] Almutairi ZM. Effect of nano-silicon application on the expression of salt tolerance genes in germinating tomato (*Solanum lycopersicum* L.) seedlings under salt stress. Plant Omics Journal. 2016;**9**(1):106-114

[32] Soliman AS, El-feky SA, Darwish E. Alleviation of salt stress on Moringa peregrine using foliar application of nanofertilizers. Journal of Horticulture and Forestry. 2015;7(2):36-47

[33] Torabian S, Zahedi M. Khoshgoftarmanesh effect of foliar spray of zinc oxide on some antioxidant enzymes activity of sunflower under salt stress. Journal of Agricultural Science and Technology. 2016;**18**:1013-1025

[34] Rezvani N, Sorooshzadeh A, Farhadi N. Effect of Nano-silver on growth of saffron in flooding stress world academy of science, engineering and technology. International Journal of Biological, Biomolecular, Agricultural, Food and Biotechnological Engineering. 2012;**6**(1):10-16

[35] Mustafa G, Sakata K, Komatsu S. Proteomic analysis of flooded soybean root exposed to aluminum oxide nanoparticles. Journal of Proteomics. 2015;**128**:280-297

[36] Hattori T, Inanaga S, Araki H, An P, Morita S, Luxová M, et al. Application of silicon enhanced drought tolerance in *Sorghum bicolour*. Physiologia Plantarum. 2005;**123**:459-466

[37] Zhu J, Wei G, Li J, Qian Q, Yu J. Silicon alleviates salt stress and increases antioxidant enzymes activity in leaves of salt-stressed cucumber (*Cucumis sativus* L.). Plant Science. 2004;**167**:527-533

[38] Cakmak I, Yilmaz A, Torun B, Erenoglu B, Broun HJ. Zinc deficiency as a critical nutritional problem in wheat production in Central Anatolia. Plant & Soil. 1996;**180**:165-172

[39] Zheng L, Hong F, Lu S, Liu C. Effect of Nano-TiO_2 on strength of naturally aged seeds and growth of spinach. Biological Trace Element Research. 2005;**104**(1):83-91

[40] Hong F, Zhou J, Liu C, Yang F, Wu C, Zheng L, et al. Effect of nano-TiO_2 on photochemical reaction of chloroplasts of spinach. Biological Trace Element Research. 2005;**105**:269-279

[41] Gao F, Hong F, Liu C, Zheng L, Su M, Wu X, et al. Mechanism of nano-anatase TiO_2 on promoting photosynthetic carbon reaction of spinach. Biological Trace Element Research. 2006;**111**:239-253

[42] Lei Z, Mingyu S, Chao L, Liang C, Hao H, Xiao W, et al. Effects of nanoanatase TiO_2 on photosynthesis of spinach chloroplasts under different light illumination. Biological Trace Element Research. 2007;**119**:68-76

[43] Yang F, Hong F, You W, Liu C, Gao F, Wu C, et al. Influence of nano-anatase TiO_2 on the nitrogen metabolism of growing spinach. Biological Trace Element Research. 2006;**110**(2):179-190

[44] Sedghi M, Hadi M, Toluie SG. Effect of nano zinc oxide on the germination parameters of soybean seeds under drought stress. Annals of West University of Timisoara Series of Biology. 2013;**XVI**:73-78

[45] Davar F, Zareii AR, Amir H. Evaluation the effect of water stress and foliar application of Fe nanoparticles on yield, yield components and oil percentage of safflower (*Carthamus tinctorious* L.). International Journal of Advanced Biological and Biomedical Research. 2014;**2**(4):1150-1159

[46] Hojjat. The effect of silver nanoparticle on lentil seed germination under drought stress. International Journal of Farming and Allied Sciences. 2016;**5**(3):208-212

[47] Ashraf M. Organic substances responsible for salt tolerance in *Eruca sativa*. Biologia Plantarum. 1994;**36**:255-259

[48] Sabaghnia N, Janmohammad M. Effect of nano-silicon particles

application on salinity tolerance in early growth of some lentil genotypes. Annales UMCS, Biologia. 2015;**69**(2):39-55

[49] Raven JA. Transport and function of silicon in plants. Biological Reviews. 1982;**58**:179-207

[50] Yang L, Watts DJ. Particles surface characteristics may play an important role in phytotoxicity of alumina nanoparticles. Toxicology Letters. 2005;**158**:122-132

[51] Sharma P, Bhatt D, Zaidi MGH, Saradhi PP, Khanna PK, Arora S. Silver nanoparticle mediated enhancement in growth and antioxidant status of *Brassica juncea*. Applied Biochemistry and Biotechnology. 2012;**167**:2225-2233

[52] Jiang HS, Li M, Chang FY, Li W, Yin LY. Physiological analysis of silver nanoparticles and $AgNO_3$ toxicity to *Spirodela polyrhiza*. Environmental Toxicology and Chemistry. 2012;**31**:1880-1886

[53] Gunjan B, Zaidi MGH, Sandeep A. Impact of gold nanoparticles on physiological and biochemical characteristics of *Brassica juncea*. Journal of Plant Biochemistry and Physiology. 2014;**2**:3, vl2

[54] Mazumdar H, Ahmed GU. Phytotoxicity effect of silver nanoparticles on *Oryza sativa*. International Journal of ChemTech Research. 2011;**3**:1494-1500

[55] Kaveh R, Li YS, Ranjbar S, Tehrani R, Brueck CL, Van Aken B. Changes in *Arabidopsis thaliana* gene expression in response to silver nanoparticles and silver ions. Environmental Science & Technology. 2013;**47**:10637-10644

[56] Mirzajani F, Askari H, Hamzelou S, Schober Y, Römpp A, Ghassempour A, et al. Proteomics study of silver nanoparticles toxicity on *Oryza sativa* L. Ecotoxicology and Environmental Safety. 2014;**108**:335-339

[57] Landa P, Vankova R, Andrlova J, Hodek J, Marsik P, Storchova H, et al. Nanoparticle-specific changes in *Arabidopsis thaliana* gene expression after exposure to ZnO, TiO_2, and fullerene soot. Journal of Hazardous Materials. 2012;**241-242**:55-62. DOI: 10.1016/j.jhazmat.2012.08.059

[58] Khodakovskaya M, Dervishi E, Mahmood M, Yang X, Li Z, Fumiya W, et al. Carbon nanotubes are able to penetrate plant seed coat and dramatically affect seed germination and plant growth. American Chemical Society Nano. 2009;**3**:3221-3227

Chapter 7

Superoxide Dismutase: A Stable Biochemical Marker for Abiotic Stress Tolerance in Higher Plants

Mukesh K. Berwal and Chet Ram

Abstract

Superoxide dismutases (SODs) are ubiquitous metalloenzymes that constitute the first line of defense against reactive oxygen species (ROS). It constitutes one of the major enzymatic components of detoxification of superoxide radicals generated in biological system by catalyzing its dismutation to H_2O_2 and finally to H_2O and O_2 by catalase and peroxidase. Most plant species contain numerous SOD isoforms differing in their active site metal ions, namely Cu/Zn-SOD, Mn-SOD, and Fe-SOD. Many studies also reported that the tolerance level of plants is positively correlated with SOD activity as well as with the number of SOD isoforms, and established the fact that "More the SOD Activity, More the Stress Tolerance." Therefore, the SOD isozyme profile of any plant can be used as stable marker for stress tolerance in plant. In this chapter, we have discussed the role of SOD in abiotic stress tolerance, type of SODs, and correlation of its activity and its isoforms with stress tolerance level.

Keywords: superoxide dismutase, isoforms, ROS, stress tolerance

1. Introduction

Plants are sessile in nature and, as a result, they do not have the capability to escape from the site of unfavorable environment. As per circumstances, plants often face the challenges to grow under adverse environmental conditions such as water deficit or excess, high intense light, low or high temperature, salinity, heavy metals, UV rays, insect and pests attack, etc. These stresses wield adverse effects on plant growth and development by inducing many metabolic changes, such as the occurrence of an oxidative stress [1–3]. As a principal cause of global crop failure, abiotic stresses decrease average yields for major crops by more than 50% [4]. Abiotic stresses impact on growth, development and productivity, and significantly limit the global agricultural productivity mainly by impairing cellular physiology/biochemistry via elevating reactive oxygen species (ROS) generation. The production of ROS during abiotic stresses results from pathways such as photorespiration, the photosynthetic apparatus, and mitochondrial respiration. Additionally, pathogens and wounding or drought or osmotic stress have been also shown to activate the production of ROS by NADPH oxidases [5–8]. The enhanced production of reactive oxygen species (ROS) during stress can pose a threat to cells, but it is also thought that ROI act as signals for the activation of stress-response and defense pathways

[9, 10]. Thus, ROS can be viewed as cellular indicators of stress and as secondary messengers involved in the stress-response signal transduction pathway.

However, several anabolic and catabolic processes like photosynthesis and respiration occur as part of common aerobic metabolism. It has been proved that ROS are generated in different cellular compartments as mitochondria, chloroplasts, peroxisomes, cytoplasm or in the extracellular space, known as apoplast by action of different enzymes [11, 12]. In vegetative tissues, approximately 1–2% of total molecular oxygen consumption drives to the creation of ROS in normal conditions. This percentage increases when plants are subjected to stress conditions such as salinity, drought, cold stress, or high temperatures. ROS are the species generated through the reduction of molecular oxygen (O_2) that includes some free radicals such as superoxide ($O_2^{\bullet -}$), hydroxyl radical ($OH^{\bullet}$), alkoxyl ($RO^{\bullet}$), and peroxyl ($ROO^{\bullet}$), and nonradical products like hydrogen peroxide (H_2O_2) and singlet oxygen ($^1O_2$), etc. [11–13]. ROS generation is an unavoidable part and by-product in various metabolic processes, where 240 μM s^{-1} $O_2^{\bullet -}$ and 0.5 μM H_2O_2 can be observed in plants under optimal growth conditions. Further, abiotic stresses may significantly enhance the generation of varied ROS (and their reaction products) in plant cells, where stressed cells may exhibit accelerated ROS generation up to 720 μM s^{-1} $O2^{\bullet -}$ and 5–15 μM H_2O_2 [14, 15] (**Figure 1**).

Plants have lot of antioxidant systems that protect them against these potential cytotoxic effects. Antioxidant enzymes are the most important components in the scavenging system of ROS. Major nonenzymatic antioxidants include ascorbic acid (AsA), glutathione (GSH), phenolic compounds, alkaloids, nonproteinaceous amino acids, and α-tocopherols. Alternatively, the battery of enzymatic antioxidants includes ascorbate peroxidase (APX), superoxide dismutase (SOD), glutathione reductase (GR), catalase (CAT), monodehydroascorbate reductase (MDHAR), dehydroascorbate reductase (DHAR), peroxidase (POX), glutathione peroxidase (GPX), guaiacol peroxidase (GOPX), and glutathione-S-transferase (GST) [15]. Considering the major enzymatic antioxidants, SOD (EC 1.15.1.1) is ubiquitous metalloenzymes [16, 17] that constitute the first line of defense against reactive oxygen species (ROS). In living cells, SODs catalyze the dismutation of the superoxide radicals ($O_2^{\cdot -}$) into hydrogen peroxide (H_2O_2) and oxygen (O_2) and play an important role in protecting the cells against the toxic effect of superoxide radicals produced in different cell compartments [18]. In plants, the role of SOD during environmental adversity has received much attention since reactive oxygen species have been found to be produced during many stress conditions.

$$2O_2^- + 2H^- \rightarrow O_2 + H_2O_2$$

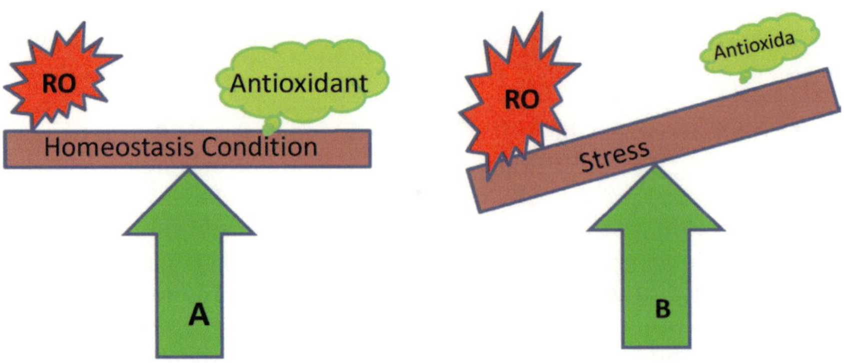

Figure 1.
The concept of homeostasis condition (A) and imbalance (B) between reactive oxygen species (ROS) and antioxidants.

2. Superoxide dismutase (SOD)

Superoxide dismutases (SODs: EC 1.15.1.1) are ubiquitous metalloenzymes [16, 17] that constitute the first line of defense against reactive oxygen species (ROS) and one of the most effective components of the antioxidant defense system in plant cells against ROS toxicity. Until reported in plants [19], SOD was recognized as a group of metalloproteins having no known function. Based on the metal cofactor at active site, SODs are categorized into three main groups and are believed to present in all oxygen-metabolizing cells and are also in all subcellular compartments like mitochondria, chloroplasts, nuclei, cytoplasm, peroxisomes, and apoplasts, etc. [20, 21]. It constitutes one of the major enzymatic components to detoxify superoxide radicals by catalyzing its dismutation to H_2O_2 [22]. By removing $O2^{•-}$, SODs decrease the risk of $OH^•$ formation via the metal catalyzed Haber-Weiss-type reaction because this reaction has a 10,000-fold faster rate than the spontaneous dismutation [11]. This enzyme is unique that its activity determines the concentrations of $O2^{•-}$ and H_2O_2, the two Haber-Weiss reaction substrates, and it is therefore likely to be central in the antioxidant defense mechanism [23, 24]. The SOD system of higher plants exhibited into multiple isoforms, which are developmentally regulated and are highly reactive against exogenous stimuli. The significance in the efficiency of all SODs has been confirmed in the direct or indirect metabolism of diverse ROS and its reaction products in numerous studies [11, 19, 25]. According to the active site metal, the multiple SOD isoforms are classified into three major groups (types): Fe-SOD (iron cofactor), MnSOD (manganese cofactor), and Cu/ZnSOD (copper and zinc as cofactors; copper is the redox active catalytic metal). While in bacteria, another type of SOD called nickel SODs (Ni-SODs) has also been reported by many researchers with nickel as metal cofactor [19, 26–28]. These multiple SOD isoforms are designated to specific cell compartments namely Fe-SODs are located in plastids, Mn-SODs in mitochondrial matrix and peroxisomes, and they also have been found in cell wall, while Cu/Zn-SODs occur in cytosol, peroxisomes, plastids, and possibly extracellular space [19, 29–31]. All SODs are encoded by nuclear genes and targeted to their respective subcellular localization by an amino terminal guiding sequence (**Table 1**).

2.1 Superoxide dismutase in plants as stable marker for abiotic stress tolerance

Different types of environmental stresses such as heat, cold, drought, salinity, and chemical contaminants commonly result in enhanced production of reactive

SOD isozymes	Structure	Subcellular localization	Sensitivity
Cu/Zn-SOD	Homodimeric and homotetrameric	Cytosol, chloroplast, peroxisome, mitochondria	H_2O_2 and KCN
Mn-SOD	Homodimeric and homotetrameric	Mitochondria, peroxisome	$CHCl_3$:CH_3CH_2OH but not to H_2O_2 and KCN
Fe-SOD	Homodimeric and tetrameric	Cytosol, chloroplast, peroxisome, Mitochondria	H_2O_2 but not to KCN
Ni-SOD	Only reported in prokaryotes		

Table 1.
Types of plant SOD, subcellular location, and sensitivity.

oxygen species (ROS), and demand an effective scavenging system to prevent oxidative damage to living cells under such conditions. Thus, the understanding of the plant responses to these abiotic stresses has become a prerequisite in order to develop crop plants tolerating abiotic stresses. Nevertheless, as an important component of plant defense machinery within a cell, SODs are major enzymatic components of the cellular defense system against abiotic stress-accrued enhanced ROS. SODs are ubiquitous to aerobic organism and catalyze the dismutation of superoxide to molecular oxygen and hydrogen peroxide (H_2O_2). Under normal conditions, the resulting H_2O_2 is effectively scavenged by catalase and peroxidase enzymes. Hereunder, recent reports available on the modulation of SODs in abiotic-stressed plant species are discussed.

It has been observed under numerous studies that the higher the SOD activity or higher number of isoforms, greater the potential to remove ROS. The upregulation of SODs is implicated in combating over-produced ROS due to biotic or abiotic stresses and has a crucial role in the survival of the plant under stressful environment. Significant increase in total leaf SOD activities as well as some extra SOD isoforms (in some studies) has been reported in many plant species under various types of abiotic stresses, namely drought, salt, and heavy metals (Cu, Cd, etc.), in a number of crops like *Arabidopsis*, mulberry [32], tomato [33], *Brassica juncea* [34, 35], *Treaticum aestivum* [36], *Hordeum vulgare, Vigna mungo* [37], citrus [38], etc. The abundance of SOD transcripts is observed in response to various abiotic and biotic stresses to distinct the oxidative stress that exerts a significant role in stress tolerance. Over-expressing transgenic plants of various SOD isoforms increases enhanced tolerance to oxidative stress and to other environmental stresses. These results have been reported in many crops and model species including *Arabidopsis,* alfalfa, rice, potato, poplar, and tobacco [39]. There have been many reports in the development of stress tolerant plants with increased expression of different SODs namely over-expressed Mn-SOD in GM *Arabidopsis* [40] and tomato [41] exhibited higher tolerance to salt, Cu/Zn-SOD overexpression in tobacco plant exhibited tolerance toward multiple stresses [42]. Furthermore, Lee et al. [43] reported that combined overexpression of Cu/Zn-SOD and ascorbate peroxidase in GM *Festuca arundinacea* plant exhibited multiple tolerance against drought (Methyl vilogen), H_2O_2, Cu, and Cd.

Berwal et al. [44] studied the SOD isozymes pattern of 13 coconut genotypes comprising six tall, five dwarfs along with two reciprocal hybrids of WCT (tall) with COD (dwarf). Among the genotypes studied, a significant variation was observed in SOD enzyme activity as well as in SOD isoforms pattern. A total of 8–14 SOD isoforms were detected in different coconut cultivars (**Figures 2** and **3**). The variation was observed only in Mn-SOD isoforms, while Fe-SOD (two) and Cu/Zn-SOD (five) isoforms were similar in all the analyzed cultivars; these isoforms have already been identified in coconut by Kumar et al. [25]. Mn-SOD isoforms varied from one to five in numbers. Among the tall cultivars, WCT, FMST, and WCT X COD showed highest number (five) of Mn-SOD isoforms as well as highest enzymatic activity followed by LCT while TPT, PHOT, and ADOT showed only single isoform for Mn-SOD. All dwarfs studies showed that they had similar SOD isozyme profile for all SODs, that is, one Mn-SOD, five Cu/Zn-SOD, and two Fe-SOD isoforms. They also observed that Mn-SOD does not follow the Mendelian pattern of inheritance, that is, reciprocal crosses showed Mn-SOD isoform pattern similar to their mother palm.

Rajgopal et al. [45] also studied the tolerance level of different coconut cultivars including the abovementioned cultivars on the basis of some physiological parameters like stomatal conductance, leaf water potential, and epicuticular wax content and scored them with 1–20 rank and WCT X WCT and FMST secured first and second ranks, respectively. Since, Berwal et al. [44] reported maximum SOD isoforms in WCT and FMST cultivars and the same are already reported as

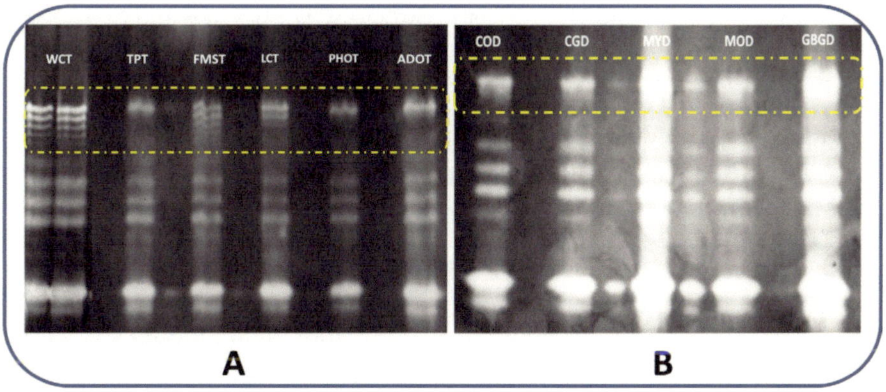

Figure 2.
Manganese superoxide dismutase (Mn-SOD) isoform variability in coconut genotypes (circled): (A) tall accessions and (B) dwarf accessions [44].

highest stress tolerant cultivars. Similarly, Kumar et al. [38] evaluated basal enzymatic antioxidative metabolism in the developing leaves of commercially grown citrus such as grapefruit, Hamlin (sweet orange), and kumquat. Young leaves of kumquat exhibited lower rates of lipid peroxidation and H_2O_2 generation as compared to grapefruit and sweet Hamlin. The total superoxide dismutase (SOD) activity, which catalyzes the transmutation of superoxide ion to H_2O_2, was two-fold higher in kumquat than grapefruit and sweet orange. Kumquat also showed more superoxide dismutase isoforms activities (**Figure 1**. Isoforms of superoxide dismutase (SOD; Panel A) and band intensities (Panel B) in developing leaves of different genotypes of citrus and kumquat at pp. 93, Kumar et al. [38]).

Despite the higher superoxide dismutase activity in kumquat, it had substantially lower H_2O_2 than grapefruit and Hamlin; and this is well-known that kumquat has greater resistance towards oxidative stresses. Gueta-Dahan et al. [46] also reported in citrus, callus, and cold-acclimated mandarin fruits and suggested higher SOD activity conferred greater resistance to salt and chilling stress (Figure 8. SOD activities in salt-sensitive (L) and salt-tolerant (R) citrus cells at pp. 465). Vysniauskiene et al. [47] reported higher SOD activity in frost-resistant potato hybrids than that of in frost-sensitive *Solanum tuberosum* "Matilda."

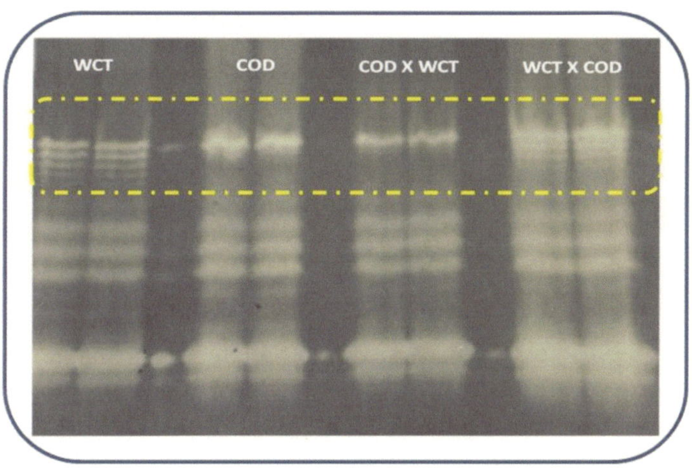

Figure 3.
Manganese superoxide dismutase (Mn-SOD) isoform pattern of WCT, COD, and their reciprocal crosses [44].

Activities of cytosolic and chloroplastic Cu/Zn-SOD isozymes and cytosolic APX (cAPX), as well as their corresponding mRNA transcripts, were increased by drought treatment of pea plants [48]. Similarly, osmotic stress increased the Mn-SOD transcript abundance in maize [49]. The higher level of gene expression corresponding to this isozyme as well as for Cu/Zn-SOD, were also increased by chilling stress in tobacco plants [50]. It has been reported in many studies that higher level of Mn-SOD is linked with abiotic stress tolerance and Melchiorre et al. [51] reported photo-oxidative stress tolerance, lower oxidative damage, and higher H_2O_2 in *Triticum aestivum* plant transformed with Mn-SOD gene from *Nicotiana plumbaginifolia*. Wang et al. [40] also reported overexpression of Mn-SOD gene in *Arabidopsis* leads to salt tolerance. Similarly, Rubico et al. [52] reported mild water stress tolerance and higher photosynthetic activity in *Medicago sativa* L. plants transformed with Mn-SOD and Fe-SOD from *Nicotiana plumbaginifolia* and *Arabidopsis thaliana*.

3. Conclusion

Superoxide dismutase is known as the first line of defense against oxidative stresses in plants and play most vital role is scavenging the reactive oxygen species produced during metabolic processes as well as under abiotic stress conditions. From the above discussion, it is clear that the plant has more native or induced SOD activity that showed more tolerance toward abiotic stresses. Many studies have proved that higher the native SOD activity along with more number of SOD isoforms makes the plants more capable to scavenge the ROS generated during stressed condition more effectively. As reported by Berwal et al. [44] in coconut and Kumar et al. [38] in citrus species that the cultivar having more number of native SOD isoforms showed more tolerance against drought stress. Therefore, the native SOD isozyme profile can be used as a stable biochemical marker for screening of crop germplasm for abiotic stress tolerance.

Author details

Mukesh K. Berwal* and Chet Ram
ICAR-Central Institute for Arid Horticulture, Bikaner, Rajasthan, India

*Address all correspondence to: mkbiochem@gmail.com

References

[1] Diaz-Vivancos P, Clemente-Moreno MJ, Rubio M, Olmos E, Garcia JA, Martínez-Gómez P, et al. Alteration in the chloroplastic metabolism leads to ROS accumulation in pea plants in response to plum pox virus. Journal of Experimental Botany. 2008;**59**:2147-2160

[2] Hernandez JA, Rubico M, Olmos E, Ros-Barcelo A, Martinez-Gimez P, et al. Oxidative stress induced by long-term plum pox virus infection in peach (*Prunus persica* L. cv. GF305). Physiologia Plantarum. 2004;**122**:486-495

[3] Hernandez JA, Escobar C, Creissen G, Mullineaux P. Role of hydrogen peroxide and the redox state of ascorbate in the induction of antioxidant enzymes in pea leaves under excess light stress. Functional Plant Biology. 2004;**31**:359-368

[4] Tuteja N, Tiburcio AF, Fortes AM, Bartels D. Plant abiotic stress: Introduction to PSB special issue. Plant Signaling & Behavior. 2011;**6**:173-174

[5] Hammond-Kosack KE, Jones JDG. Resistance gene-dependent plant defense responses. The Plant Cell. 1996;**8**:1773-1791

[6] Orozco-Cardenas M, Ryan CA. Hydrogen peroxide is generated systemically in plant leaves by wounding and systemin via the octadecanoid pathway. Proceedings of the National Academy of Sciences United States of America. 1999;**96**:6553-6557

[7] Cazale AC, Droillard MJ, Wilson C, Heberle-Bors E, Barbier-Brygoo H, Lauriere C. MAP kinase activation by hypo-osmotic stress of tobacco cell suspensions: Towards the oxidative burst response? The Plant Journal. 1999;**19**:297-307

[8] Pei ZM, Murata Y, Benning G, Thomine S, Klüsener B, Allen GJ, et al. Calcium channels activated by hydrogen peroxide mediate abscisic acid signaling in guard cells. Nature. 2000;**406**:731-734

[9] Desikan R, A-H-Mackerness S, Hancock JT, Neill SJ. Regulation of the Arabidopsis transcriptosome by oxidative stress. Plant Physiology. 2001;**127**:159-172

[10] Knight H, Knight MR. Abiotic stress signaling pathways: Specificity and cross-talk. Trends in Plant Science. 2001;**6**:262-267

[11] Gill SS, Tuteja N. Reactive oxygen species and antioxidant machinery in abiotic stress tolerance in crop plants. Plant Physiology and Biochemistry. 2010;**48**:909-930

[12] Sandalio LM, Rodriguez-Serrano M, Romero-Puertas MC, del-Rıo LA. Role of peroxisomes as a source of reactive oxygen species (ROS) signaling molecules. Subcellular Biochemistry. 2013;**69**:231-255

[13] Halliwell B, Gutteridge JMC. Free Radicals in Biology and Medicine. Oxford: Oxford University Press; 2007

[14] Mittler R. Oxidative Stress in Plants. 2011. http://biology.unt.edu/ros/pages/rosmetabolismdoc

[15] Hasanuzzaman M, Hossain MA, daSilva JAT, Fujita M. Plant responses and tolerance to abiotic oxidative stress: Antioxidant defense is a key factor. In: Bandi V, Shanker AK, Shanker C, Mandapaka M, editors. Crop Stress and its Management: Perspectives and Strategies. Berlin: Springer; 2012. pp. 261-316

[16] Fridovich I. Superoxide dismutase. Annual Review of Biochemistry. 1975;**44**:147-159

[17] Jackson CA, Moore AL, Halliwell B, Foyer CH, Hall DO. Subcellular localization and identification of superoxide dismutase in the leaves of higher plants. European Journal of Biochemistry. 1978;**91**:339-344

[18] Del Rio LA, Sandalio LM, Altomare DA, Zilinskas BA. Mitochondrial and peroxisomal manganese superoxide dismutase: Differential expression during leaf senescence. Journal of Experimental Botany. 2003;**54**:923-933

[19] McCords JM, Fridovich I. Superoxide dismutase. The Journal of Biological Chemistry. 1969;**44**(22):6049-6055

[20] Alscher RG, Erturk N, Heath LS. Role of superoxide dismutases (SODs) in controlling oxidative stress in plants. Journal of Experimental Botany. 2002;**53**(372):1331-1341

[21] Fink RC, Scandalios JG. Molecular evolution and structure—function relationships of the superoxide dismutase gene families in angiosperms and their relationship to other eukaryotic and prokaryotic superoxide dismutases. Archives of Biochemistry and Biophysics. 2002;**399**(1):19-36

[22] Hossain Z, Lopez-Climent MF, Arbona V, Perez-Clemente RM, Gomez Cadenas A. Modulation of the antioxidant system in citrus under waterlogging and subsequent drainage. Journal of Plant Physiology. 2009;**166**:1391-1404

[23] Bowler C, Van Montagu M, Inze D. Superoxide dismutase and stress tolerance. Annual Review of Plant Physiology and Plant Molecular Biology. 1992;**43**:83-116

[24] Kandhari P. Generic differences in antioxidant concentration in the fruit tissues of four major cultivars of apples [Master thesis]. College Park: University of Maryland; 2004

[25] Kumar M, Sugatha P, Hebbar KB. Superoxide dismutase isozymes and their heat stability in coconut (Cocus nucefera L.) leaves. Annals of Biology. 2014;**30**(4):593-597

[26] Reddy CD, Venkaiah B. Isoenzymes of superoxide dismutase from mung bean (Phaseolus aureus) seedlings. Current Science. 1982;**20**:987-988

[27] Werner K, Heinz R, Andrea P. Purification of two SOD isozymes and their sub-cellular localization in needles and roots of Norway spruce (*Picea abies* L.). Plant Physiology. 1992;**100**:334-340

[28] Pradedova EV, Isheeva OD, Salyaev RK. Superoxide dismutase of plant cell vacuoles. Membrane and Cell Biology. 2009;**26**(1):21-30

[29] Kukavica B, Mojovi M, Vucinic Z, Maksimovic V, Takahama U, Veljovic-Jovanovic S. Generation of hydroxyl radical in isolated pea root cell wall, and the role of cell wall-bound peroxidase, Mn-SOD and phenolics in their production. Plant and Cell Physiology. 2009;**50**:304-317

[30] Pilon M, Ravet K, Tapken W. The biogenesis and physiological function of chloroplast superoxide dismutases. Biochimica et Biophysica Acta. 2011;**1807**:989-998

[31] Miller AF. Superoxide dismutases: Ancient enzymes and new insights. FEBS Letters. 2012;**586**:585-595

[32] Harinasut P, Poonsopa D, Roengmongkol K, Charoensataporn R. Salinity effects on antioxidant enzymes in mulberry cultivar. Science Asia. 2003;**29**:109-113

[33] Gapinska M, Sklodowska M, Gabara B. Effect of short- and long-term salinity on the activities of antioxidative enzymes and lipid peroxidation in tomato roots. Acta Physiologiae Plantarum. 2008;**30**(1):11-18

[34] Mobin M, Khan NA. Photosynthetic activity, pigment composition and antioxidative response of two mustard (*Brassica juncea*) cultivars differing in photosynthetic capacity subjected to cadmium stress. Journal of Plant Physiology. 2007;**164**:601-610

[35] Li Y, Song Y, Shi G, Wang J, Hou X. Response of antioxidant activity to excess copper in two cultivars of Brassica campestris ssp. chinensis Makino. Acta Physiologiae Plantarum. 2009;**31**:155-162

[36] Khan NA, Samiullah S, Nazar R. Activities of antioxidative enzymes, Sulphur assimilation, photosynthetic activity and growth of wheat (*Triticum aestivum*) cultivars differing in yield potential under cadmium stress. Journal of Agronomy and Crop Science. 2007;**193**:435-444

[37] Guo T, Zhang G, Zhou M, Wu F, Chen J. Effects of aluminium and cadmium toxicity on growth and antioxidant enzyme activities of two barley genotypes with different Al resistance. Plant and Soil. 2004;**258**:241-248

[38] Kumar N, Ebel RC, Roberts PD. Antioxidant isozyme variability in different genotypes of citrus and kumquat. Journal of Crop Improvement. 2011;**25**(1):86-100. DOI: 10.1080/15427528.2011.545280

[39] Van Camp W, Willekens H, Bowler C, Van Montagu M, Inze D, Reupold-Popp P, et al. Elevated levels of superoxide dismutase protect transgenic plants against ozone damage. Nature Biotechnology. 1994;**12**:165-168

[40] Wang Y, Ying Y, Chen J, Wang XC. Transgenic Arabidopsis overexpressing Mn-SOD enhanced salt-tolerance. Plant Science. 2004;**167**:671-677

[41] Wang Y, Wisniewski M, Meilan R, Uratsu SL, Cui MG, Dandekar A, et al. Ectopic expression of Mn-SOD in Lycopersicon esculentum leads to enhanced tolerance to salt and oxidative stress. Journal of Applied Horticulture. 2007;**9**:3-8

[42] Badawi GH, Yamauchi Y, Shimada E, Sasaki R, Kawano N, Tanaka K, et al. Enhanced tolerance to salt stress and water deficit by over expressing superoxide dismutase in tobacco (*Nicotiana tabacum*) chloroplasts. Plant Science. 2004;**166**:919-928

[43] Lee SH, Ahsan N, Lee KW, Kim DH, Lee DG, Kwak SS, et al. Simultaneous over expression of both Cu/Zn superoxide dismutase and ascorbate peroxidase in transgenic tall fescue plants confers increased tolerance to a wide range of abiotic stresses. Journal of Plant Physiology. 2007;**164**:1626-1638

[44] Berwal MK, Sugatha P, Niral V, Hebbar KB. Variability in superoxide dismutase isoforms in tall and dwarf cultivars of coconut (*Cocos nucifera* L.) leaves. Indian Journal of Agriculture Biochemistry. 2016;**29**(2):184-188

[45] Rajgopal V, Kastiribai KV, Voleti SR. Screening of coconut genotypes for drought tolerance. Oleagineux. 1990;**45**:2015-2220

[46] Gueta-Dahan Y, Yaniv Z, Zilinskas BA, Ben-Hayyim G. Salt and oxidative stress: Similar and specific responses and their relation to salt tolerance in citrus. Planta. 1997;**203**(4):460-469

[47] Vysniauskiene R, Ranceliene V, Radziunaite-Paukstiene A, Spalinskas R. The UV-B impact upon the enzyme of antioxidant system superoxide dismutase (SOD) of potato somatic hybrids. Biologia. 2007;**53**:67-70

[48] Mittler R, Zilinskas BA. Regulation of cytosolic ascorbate peroxidase and other antioxidant enzymes during the progression of drought stress and

following recovery from drought. Plant Journal. 1994;**5**:397-405

[49] Zhu D, Scandalios JG. Differential accumulation of manganese-superoxide dismutase transcripts in maize in response to abscisic acid and high osmoticum. Plant Physiology. 1994;**106**:173-178

[50] Tsang EWT, Bowler C, Herouart D, Van Camp W, Villarroel R, Genetello C, et al. Differential regulation of superoxide dismutases in plants exposed to environmental stress. The Plant Cell. 1991;**3**:783-792

[51] Melchiorre M, Robert G, Thippi V, Racca R, Lascano HR. Superoxide dismutase and glutathione reductase overexpression in wheat protoplast: Photooxidation stress tolerance and change in cellular redox state. Plant Growth Regulation. 2009;**57**:57-68

[52] Rubio MC, Gonzalez EM, Minchin FR, Web KJ, Arrese-Igor C, Ramos J, et al. Effect on water stress on antioxidant enzymes of leaves and nodules of transgenic alfalfa overexpressing superoxide dismutase. Physiologia Plantarum. 2002;**115**:531-540

Chapter 8

Melatonin: Role in Increasing Plant Tolerance in Abiotic Stress Conditions

Raziye Kul, Aslıhan Esringü, Esin Dadasoglu, Üstün Sahin, Metin Turan, Selda Örs, Melek Ekinci, Guleray Agar and Ertan Yildirim

Abstract

Nowadays, due to the environmental stress factors that limit the production of crops, it has become very difficult to find suitable areas to enable the plant to reach its optimum product potential. Abiotic stress is very effective in decreasing agricultural production. Factors such as drought, salinity, high and low temperature, flood, radiation, heavy metals, oxidative stress, and nutrient deficiency can be considered as abiotic stress factors, and these sources of stress negatively affect plant growth, quality and productivity. Melatonin (MEL) was first identified in plants in 1995 and is increasingly becoming important for its role and effects in the plant system. MEL has been shown to have a substantial role in plant response to growth, reproduction, development, and different stress factors. In addition to its regulatory role, MEL also plays a protective role against different abiotic stresses such as metal toxicity, temperature, drought, and salinity. In plants, an important role of MEL is to alleviate the effects of abiotic stresses. In this review, the effects of MEL on plant growth, photosynthetic activity, metabolism, physiology, and biochemistry under abiotic stress conditions as a plant growth regulator will be examined.

Keywords: melatonin, plant development, physiology, abiotic stress

1. Introduction

Plants are exposed to various adverse conditions that limit their growth in areas where they are grown. Conditions that prevent growth, development, and metabolism in plants are called stress [1]. Due to the continuation of climate change and the increase in extreme climatic conditions, it is reported that the negative impact of environmental stress factors on plant production will increase in many regions of the world [2]. Stress factors can simultaneously show their effects on plants [3]. On the reducing amount of agricultural product, abiotic stress has reached a massive quantity of 71% although other stresses are on 29% [4]. It is estimated that only 10% of the arable land in the world is far from some forms of stress. It has been reported that abiotic stress factors are the main limiting factors of crop production in the world and cause more than 50% reduction in the yield of most products [5].

In the last decade, different irrigation techniques, soil improvement, and the use of suitable fertilizers have been intensified in order to reduce the impact of major stress factors. As a different approach, the use of some externally applied healers during plant growth has been tried in recent years, and it has been observed that the applications using MEL may have an effect of increasing stress tolerance in the plant.

MEL (N-acetyl-5-methoxytiprimamine) was discovered in 1958 in the cattle pineal gland [6]. MEL has been one of the most investigated biological molecules, which is extensively researched in animals. MEL was first explored in plants in 1995 and is an indoleamine neurohormone [7, 8]. There has been an increasing interest in MEL's roles and impacts on metabolic processes. It was found to play a major role in various plant reactions such as growth, flowering, development, and stress [9–11]. Most of the reports that provide information about these processes are based on analytical analysis to determine the internal MEL content in response to a stimulus, treatment, or mutation, whereas administration usually requires prolonged exposure or treatment of plants under in vitro or greenhouse conditions [9].

Several studies have reported that MEL can be considered a growth regulator because it plays a role in specific physiological events in plants. Indeed, MEL regulates the growth of leaf, shoots and explants, and the leaf senescence. The natural antioxidant capacity of MEL can be explained by its ability to increase tolerance in plants exposed to abiotic stresses such as drought, cold, heat, salinity, chemical pollutants, herbicides, and UV rays [12]. MEL's capability to behave as a plant biostimulator for biotic and abiotic stress conditions and the ability to regulate plant growth can regulate plant vegetative growth processes such as rooting, leaf aging, photosynthetic yield, and biomass yield, and it plays a potential regulatory role in flowering processes and the formation and maturation of fruit and seeds [10, 12, 13].

In this study, the effects of MEL on plant growth and physiology against some abiotic stress factors that have important impacts on plant growth and development have been given according to the findings of various researches.

2. MEL and its function in plant growth and physiology

MEL regulates various metabolic processes in animals and plants. MEL is an endogenously produced molecule in all plant species that have been investigated. Its concentration in plant organs varies in different tissues, e.g., roots versus leaves, and with their developmental stage [10].

MEL, tryptophan, tryptamine, and serotonin are structural biogenic indoleamine and also related to indole-3acetic acid (IAA) and indolic compounds, which are very important in plant physiology such as common auxin. Metabolic pathways of tryptophan in mammals and plants as proposed by Murch et al. [14] are shown in **Figure 1**. MEL was determined in the roots, leaves, fruits, and seeds of various plant species. Melatonin has been suggested to function as an auxin to promote vegetative growth in a number of plant species [15]. For instance, Murch et al. [16] used auxin, serotonin, and MEL inhibitors to demonstrate the role of MEL in plant growth and found that the high intrinsic MEL concentration promotes root growth in the *Hypericum perforatum* L. plant, while the increase in serotonin concentration, the precursor to MEL, promotes body development. It takes part in many different tasks in the metabolic processes in the plant. It has been shown to regularize proline metabolism [17]. It contributes root formation and water balance, thus keeping membrane and chlorophyll integrity [18, 19].

Earlier researchers have inspected the physiochemical effects of MEL on plants. These reports indicated that MEL has a role in various plant metabolic processes as the modulation of the flowering physiology and development. Furthermore, it also

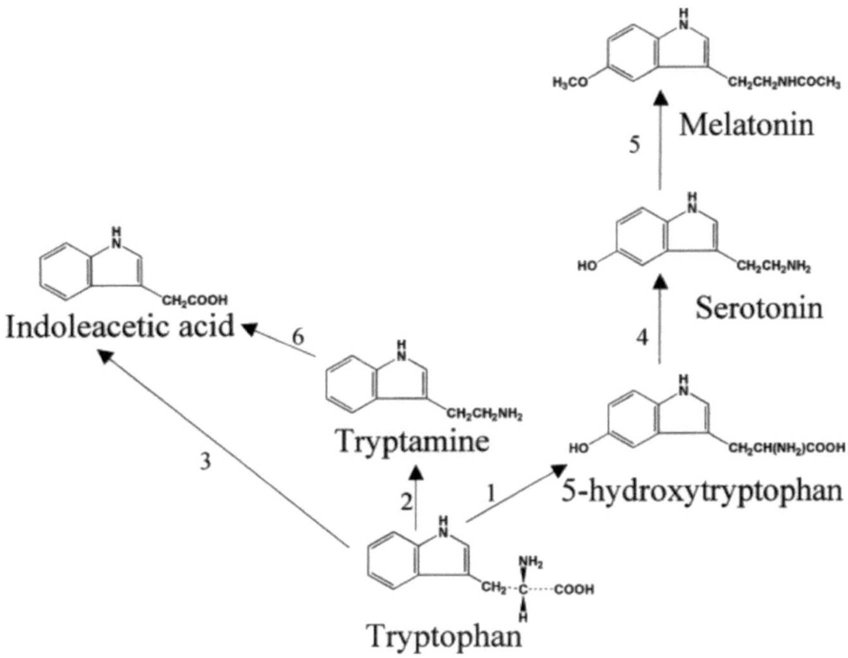

Figure 1.
Metabolic pathways of tryptophan in mammals and plants as proposed by Murch et al. [14].

postpones flower formation, countenances plant growth and biomass production, and hinders chlorophyll degradation [20, 21]. It has been shown that MEL roles in plants could be: growth promoters as auxins, antioxidants for ROS, and other roles as signal molecules [15]. It acts in various plant cellular metabolic and biological processes, including rooting [22], chlorophyll catabolism [20], and stress tolerance [23, 24]. Plants can synthesize MEL, and it plays a role as an antioxidant or a modulator of growth and development in plants [25].

Similar effects (growth induction or inhibition at high levels because of auxin-stimulated ethylene biosynthesis) of MEL were determined in the monocotyledons [26]. Furthermore, MEL applications enhanced photosynthetic capacity, redox homeostasis, and root formation in various crops [22, 27, 28]. According to another report, a coating of soybean seed with MEL notably increased plant growth and seed yield [19].

It has been reported that MEL affects lateral root formation in lupin, and this effect is very similar to the effect of IAA [29]. In these studies, auxin-induced root and cytokinin-induced shoot organogeneses were inhibited by alterations in the endogenous concentration of MEL and inhibitors of the transport of serotonin and MEL [30]. MEL has been reported to regulate seed germination, growth of roots, shoots and explants, and leaf senescence [12]. In addition, Tan et al. [31] pointed out that high MEL content in plants increased the germination rate of seeds under adverse conditions, increasing the life expectancy and improving the quality of plant production. In lupin, MEL increased plant root and shoot biomass with similar results to IAA for root biomass in concentrations used [32]. Exogenous MEL was applied to etiolated wild mustard, and the effect on root growth and endogenous indole-3-aceticacid (IAA) levels was detected in wild mustard. Exogenous lower MEL concentrations also elevated the endogenous IAA content in roots, whereas higher levels did not significantly affect the IAA content. The specific mechanism which leads exogenous MEL to increase the IAA content in roots, associated with root formation, continues to emerge [21].

Increased MEL levels in plants have been suggested to mitigate various pollutant effects via behaving as a ROS scavenger and antioxidant. MEL can detoxify the OH^-, H_2O_2, nitric oxide, ONOO−, HNO3, and HClO, which are biosynthesized under stress conditions. Moreover, MEL treatments elevate antioxidant enzyme activities under abiotic stress conditions. One MEL molecule has the ability to scavenge up to 10 free radicals [33]. MEL has also been suggested to have a significant effect on nitrogen and carbohydrate metabolism and on transcription rearrangement [34]. Thus, it is suggested that MEL affects signal transduction and also has an important role in regulating physiological and biological processes. As a conclusion, MEL could be considered as a biological growth regulator to increase the production capacity in crops.

3. Role of melatonin in mitigating abiotic stresses

3.1 Salinity and MEL applications

Extreme salinity in soil solution is the major abiotic stress factor that drastically limits crop productivity worldwide. Salinity affects 110 million hectares in arid and semi-arid regions. According to FAO, an estimated 20–30 million ha area is severely deteriorated due to salinity [34]. In addition to the natural conditions, problems with salinity have increased with the fact that water tables have increased and concentrated in a large part of the land that has been recently irrigated [35]. Moreover, use of the treated and untreated wastewater at increasing proportions due to the insufficient clean water resources in the world can promote the soil salinity. Salinity causes osmotic stress by reducing the water potential and increasing the energy required for the intake of water and nutrients. Ionic stress is caused by the accumulation of sodium and chlorine ions in sensitive plant tissues [36–38]. Therefore, it has been reported that high concentrations of salt (especially NaCl) in soils or irrigation water disrupt the morphological and physiological processes in plants and prevent growth [39].

In addition, salinity conditions may lead to nutritional disorders and deficiencies [40]. In the short term, while the water availability reduces due to inducing osmotic stress under salty conditions, in the long term, the nutrient imbalances induce ion toxicity [41]. Salinity increases ROS formation and stimulates oxidative stress [42], which causes substantial injury to membranes and other cellular structures [43]. Salt stress affects plant physiology at both plant and cellular levels by osmotic and ionic stress. High salt concentrations may adversely affect seed germination, seedling growth, vegetative growth, flowering and fruit behavior, and photosynthetic activity and ultimately reduce economic yield and quality [44].

Lately, the positive roles of MEL in plant salt stress resistance have been progressively evolved by two ways: the exogenous application of MEL or genetic modification of the enzymes involved in MEL synthesis [45]. Indeed, exogenous MEL applications improved growth, photosynthetic capacity, antioxidant activity, and chlorophyll content, but decreased the ROS level and oxidative injury in cucumber grown under salinity stress conditions [46]. Dawood et al. [47] indicated that exogenous MEL applications enhanced plant biomass, relative water content, photosynthetic activity, phenolic matter, and plant nutrient uptake, and reduced the Na and Cl content in fava bean under salinity stress conditions. 500 mM of MEL was more effective than 100 mM on observed parameters.

Zhou et al. [48] investigated influences of MEL treatments on photosynthetic activity in tomato under salinity conditions. They concluded that MEL treatments mitigated the deleterious effects of salinity on growth and photosynthetic capacity. It was determined that MEL decreased the ROS levels and expedited the recovery of the photosynthetic electron transport chain and protein biosynthesis, therefore improving photosynthetic capacity under salinity stress. Similarly, MEL treatments on roots of watermelon mitigated salt-stress damage in photosynthetic capacity and oxidative stress, improving redox homeostasis and antioxidant enzyme activity [28]. In another study, it was reported that MEL treatments improved the tolerance to salt stress and K^+/Na^+ homeostasis in potato, increasing K^+ and decreasing NaCl concentration [44].

In *Cucumis sativus*, especially after treatment with MEL under salinity conditions, an increase in seedling growth, nutrient intake, and nitrogen metabolism was observed [49]. Ke et al. [50] proved that MEL pretreatments alleviated the negative impact of salinity stress by regulating polyamine metabolism in wheat. They also suggested that MEL could induce enzyme activity, resulting in stimulating ROS scavenging antioxidant defense in response to salinity. In another study, it was concluded that 10–500 μM MEL solutions enhanced germination and seedling growth in rice under salinity conditions. This enhancement was attributed to reducing the contents of Na^+ and Cl^- in roots and leaves [51]. The results of Jiang et al. [52] showed that exogenous MEL treatments on salt stressed maize plants caused a notable improvement in growth, photosynthetic capacity, antioxidant enzyme activity, and homeostasis. It was proven that MEL concentration in roots elevated because of stress conditions, increasing to six times the MEL concentration compared to the control. This raise can play an important role in the amelioration of stress conditions [20].

Exogenous MEL treatments showed a major effect of MEL related to lipid metabolism with K^+/Na^+ homeostasis in a potato grown under salinity stress [44]. MEL applications on roots mitigated the deleterious effects of salinity on photosynthetic capacity by reducing oxidative stress, improving antioxidant enzyme activity in watermelon. This effect was attributed to the inhibition of stomatal closure and enhanced light energy absorption and electron transport in photosystem II [28].

Liang et al. [23] treated plants with MEL to determine its effect on physiological and biochemical properties in rice grown under salinity stress. The results of the study showed that MEL treatments decreased or inhibited chlorophyll damage and the transcripts of senescence-associated genes, thus improving salinity tolerance. It was also determined that MEL postponed the leaf senescence and cell death by counteracting the ROS.

3.2 Drought and MEL applications

Water scarcity has been becoming a major problem worldwide due to population growth and social and economic development. A number of countries faced to water shortage is more than 100, and approximately two thirds of the world population will be exposed to be suffering from moderate to high water stress by 2025 [53]. Increasing domestic and industrial water demand and pollution of water threatens the water used in agriculture. Therefore, drought is one of the most important agricultural problems in the world. Two-fifths of world agriculture is carried out in arid areas [54]. Studies show that in the coming years, the effect of drought will increase further and this situation will affect the negative effects of agricultural production [55]. It is reported that global climate change, in addition to the expansion of arid and semi-arid areas, will increase the duration and intensity of drought, desertification processes, salinization, and erosion [56]. It has been shown in many studies that drought has an impact on all plant growth events from plant morphology to molecular levels [57]. Drought stress causes various physiological, biochemical, and molecular responses in different plants to help them adapt to such limiting environmental conditions [58, 59]. Arid conditions have a negative effect on photosynthetic activity, cause changes in chlorophyll content and components in the cell, and damage photosynthetic parts [60]. It also inhibits photochemical activity and reduces the activity of enzymes in the Calvin circle [61].

Earlier studies pointed out that exogenous MEL treatments improved plant tolerance to water deficit stress. Increased antioxidant activity in different plants grown in drought stress has been associated with the MEL content [62]. The effect of MEL application on plant development and some biochemical properties of *Brassica napus* L. under dry conditions were determined. In the study, it was determined that exogenous MEL treatments (0.05 mmol/L) mitigated the deleterious effects of water deficit on plant growth. Moreover, MEL treatments caused decreased H_2O_2 and increased antioxidant enzyme activity and osmotic solutes [63].

It has been shown that a notable increase in photosynthetic capacity and stress-related phytohormones was associated with the endogenous MEL content under water deficit conditions. Indeed, Fleta-Soriano et al. [25] proved that MEL treatments enhanced photosystem II resulting in a preserving factor in maize under drought stress. MEL treatments helped to recover from drought stress by enhancing the Fv/Fm ratio, which could have a defensive effect in plants subjected to water deficit conditions.

Cui et al. [64] demonstrated that MEL applications alleviated the deleterious effects of drought stress in wheat by increasing antioxidant activity and decreasing ROS and membrane injury. They also showed that MEL caused a thicker epidermal cell, intact grana lamella of chloroplast and leaf structure, and higher photosynthetic activity. They explained these positive responses to MEL treatments in wheat with enhanced enzyme activity and gene expression. Moreover, Wang et al. [18] proved that MEL had an ameliorative effect on drought stress by increasing antioxidant activity. Similarly, mitigation of deleterious effects of drought stress could be attributed to its ROS scavenging functions by improving antioxidant enzyme activity and photosynthetic efficiency [65].

Ma et al. [66] showed that exogenous MEL treatments elevated ME biosynthesis gene (TDC1, SNAT1, and COMT) expression, resulting in mitigation of leaf senescence caused by water deficit in *Agrostis stolonifera*. In another study, it was determined that MEL applications in drought conditions reduced electrical leakage, decreased chlorophyll degradation, and increased photosynthetic activity in two different apple cultivars that are resistant to drought. MEL applications reduce the expression of the ABA synthetic gene (MdNCED3) and increase the expression of catabolic genes (MdCYP707A1 and MdCYP707A2), thus reducing the level of ABA under dry conditions [67]. Exogenous MEL treatments have resulted in enhanced photosynthetic capacity and water use efficiency due to increased indole acetic acid (IAA) and zeatin and decreased H_2O_2 and aminocyclopropane-1-carboxylic acid (ACC) production [68].

It was determined in plants that drought increased the expression of genes related to drought stress and decreased the production of abscisic acid (ABA), which leads to the closure of stomata [69]. In addition to reducing the effect of drought stress, MEL also helps to heal the plants after drought has occurred and water is re-fed [33, 65, 69, 70].

3.3 Heavy metal stress and MEL applications

The increment in mining, factories, and industrialization leads to the contamination of larger areas with heavy metals. It is reported that heavy metals are included in the food chain by accumulation by plants [71]. Studies on heavy metal accumulation and its effects on plants have shown that heavy metals are a potent phytotoxic and cause growth inhibition and, in some cases, death [72, 73]. Metals are the elements necessary for normal survival of plants. However, the presence of some metals in the root region has a toxic effect on the plants. These metals, which have a negative effect on growth and yield in plants, are mostly cadmium (Cd), chromium (Cr), zinc (Zn), copper (Cu), lead (Pb), and nickel (Ni) [74]. They can easily accumulate in plants and prevent plant growth and nutrient uptake [75]. The metals in question prevent the uptake of necessary minerals by making a toxic effect and by replacing the necessary minerals such as iron for the plants. Heavy metals, by activating active oxygen species in plants, cause a decrease in chlorophyll and thus photosynthesis rate. As in other stress conditions, heavy metal stress also increases the level of plant ethylene, slows down the growth of roots and shoots, reduces CO_2 fixation, and limits the transport of sugar [76]. Many researchers have reported that heavy metals stimulate ROS formation, leading to oxidative stress [77–79].

The plants exposed to heavy metals (lead, zinc, cadmium, etc.) have been shown to induce MEL biosynthesis for alleviating stress effects [80]. Tan et al. [32] pointed out that MEL treatments elevated the phytoremediation capacity of pea under copper stress. Many studies have shown that exogenous MEL treatments reduced the toxic impact of various heavy metals such as cadmium, aluminum, copper, vanadium, nickel, etc. by enhancing root growth, antioxidant activity, photosynthetic capacity, and organic acid anion exudation, reducing metal concentration, and regulating MEL biosynthesis and antioxidant-related gene expression in various crops [24, 81, 82].

Tang et al. [83] reported that foliar MEL applications improved the photosynthetic capacity of eggplant under cadmium stress. They suggested that increased MEL concentration elevated photosynthetic capacity in stressed plants, and a concentration of 150 $\mu mol{\cdot}L^{-1}$ was the best for alleviating cadmium stress. Cadmium (Cd), one of the most dangerous heavy metal pollutants, is toxic to animals and plants. A significant increase in antioxidant enzyme activity and low ROS contents were related to treatment of MEL-stimulated Cd tolerance in tomato. MEL treatments induce Cd sequestration and transfer of cadmium from cytosol to the vacuole and cell wall [84]. Similarly, MEL applications mitigated Cd-stimulated oxidative stress by increasing the levels of nonenzymatic and enzymatic antioxidants. Gu et al. [85] determined that Cd stress conditions enhanced endogenous MEL concentrations in alfalfa. It was determined in their research that exogenous MEL treatments mitigated the negative effect of Cd on plant growth by reducing Cd accumulation and reestablishing the micro RNA-mediated redox homeostasis. They suggested that MEL could regulate expression of ion-channel genes in crops against Cd stress. Moreover, Safari et al. [86] concluded that excessive boron (B) decreased photosynthesis and dry matter in pepper. However, they pointed out that exogenous 1 μM MEL treatments eliminated visible B toxicity symptoms due to B, increased nutrient uptake, photosynthetic activity, antioxidant capacity, and accumulation of carbohydrates, and decreased ROS and membrane permeability.

Zhang et al. [24] tested whether exogenous MEL treatments could mitigate aluminum induced phytotoxicity in *Glycine max*. They pointed out that the effect of MEL on Al stress was dose-dependent. While 0.1 and 1 mM doses of MEL improved root growth and reduced H_2O_2 content, 100 and 200 mM doses affected negatively. 1 mM MEL root application increased antioxidant enzyme activity under Al stress conditions. Ni et al. [87] demonstrated that Cd stimulated the expression of MEL-related genes and enhanced the endogenous MEL content in wheat. Exogenous MEL treatments mitigated Cd toxicity on plant growth and increased ascorbate peroxidase (APX) and superoxide dismutase (SOD) activity. They reported that MEL had an important role in keeping H_2O_2 homeostasis by modulating antioxidant activity. Restricted growth properties, chlorophyll and carotenoid content, photosynthetic activity, and increased lipid peroxidation were alleviated by MEL treatments in spinach via enhancing antioxidant activity and reducing ROS levels [88].

3.4 Temperature extremes and MEL applications

Plants are affected at the maximum level from the environmental temperature from seed germination to product acquisition. Plants require an optimum temperature request for every stage of growth, and this requirement may vary between species and even varieties. The temperatures below the optimum negatively affect plant growth and ultimately yield. Low and high temperatures slow the seed germination and emergence, limit the intake of water and nutrients, increase the damage of diseases, negatively affect flowering, seed and fruit formation, and finally cause death of the plant [89]. Hot climate plant species are very sensitive to low temperatures [90]. It has been reported that low temperature affects the whole metabolic system of the cell and even causes water stress [91]. It has also been reported that low temperature causes damage in cell membranes, which also affects sugars, phenolics, phospholipids, protein, and ATP [92]. Low temperature is one of the most limiting abiotic stresses for crop yield and geographical distribution in plants [93, 94]. Low temperature stimulates the overproduction of ROSs in plant cells such as superoxide radical ($O2.^{-}$), H_2O_2, and hydroxyl radical (OH^{-}). ROS may lead to lipid peroxidation and oxidative modifications in proteins and nucleic acids [95, 96].
However, the plants have developed a specific protective mechanism to alleviate and repair damage induced by oxidative stress. The most important oxidative stress cleaning mechanisms are enzymatic systems consisting of SOD, POD, CAT, APX, and GR and nonenzymatic acetyl salicylic acid and glutathione (GSH) [97, 98]. Tolerance to low temperature in plants is positively related to activation of ROS cleaning systems. Research has shown that antioxidant activity has a substantial role in preserving plants to oxidative injury caused via stress [93].

High temperature stress is one of the most harmful stress conditions that damage the growth and yield of cool season plants. High temperature can negatively affect germination and output in many plant species. In the vegetative development period, it was reported that high temperature decreases photosynthesis capacity, CO_2 assimilation, and metabolic processes [99, 100]. High temperatures can also deteriorate membrane stability, resulting in necrotic spots similar to water stress symptoms in leaves, eventually leading to premature deaths [101]. Temperature stress negatively affects the food intake in plants [102]. In the generative development period, high temperature, flower dust germination, fertilization, flowering, and seed and fruit formation can cause a significant decrease in yields [103, 104]. Like other stresses, high temperature stress also has a significant negative impact on product yield. High temperature causes oxidative stress, lipid peroxidation, membrane damage, protein degradation, enzyme inactivation, pigment bleaching, and degradation of DNA strands in plants [105].

Temperature extremes were shown to increase MEL biosynthesis. Moreover, it has been reported that exogenous MEL treatments helped to protect plants from temperature extremes [106]. Several studies indicated that antioxidant capacity of MEL could strengthen plants subjected to abiotic stresses such as cold and heat [107, 108]. There are reports showing that supplementation with MEL induced MEL biosynthesis and upregulated genes under cold stress conditions [109, 110]. Studies have also shown that MEL treatments alleviated the deleterious impact on plants by upregulating or downregulating genes and proteins related to high or low temperature stresses, scavenging ROS, modulating polyamine metabolism, increasing chlorophyll and heat shock protein synthesis, and affecting the ABA and cytokinin pathway [33, 110–113].

Lei et al. [114] suggested that MEL enhanced carrot cell survival due to induced putrescine and spermidine biosynthesis under cold stress. Similarly, Balabusta et al.

[115] determined that osmo-primed cucumber seeds with MEL had lower ROS levels and higher superoxide dismutase (SOD) activity, detoxifying ROS under chilling stress. It is evidenced that exogenous MEL treatments reduced photoinhibition by enhancing nonphotochemical quenching via induction of violaxanthin de-epoxidase activity in tomato plants under chilling stress [116]. Alam et al. [117] concluded that MEL-treated tall fescue plants under high temperature stress had lower ROS electrolyte leakage and malondialdehyde levels and higher chlorophyll, total protein, and antioxidant enzyme activities compared to nontreated plants. They also showed that exogenous MEL treatments improved thermo-tolerance.

In another study, maize seeds were primed with MEL (50 and 500 μM) to determine the priming-induced changes under chilling stress. Priming with MEL regulated MEL-associated proteins in seeds exposed to lower temperature and enhanced plant tolerance to chilling [118]. Foliar MEL-treated *Lolium perenne* plants had greater biomass, chlorophyll content, and photosynthetic capacity compared to nontreated ones under heat stress. MEL also caused increased endogenous MEL and reduced ABA content. Genes related to ABA were downregulated by MEL treatments [113].

MEL applications decreased the H_2O_2 and MDA content of pepper seedlings, but increased the SOD and CAT enzyme activities in pepper under chilling stress. The decrease in the peroxidation of lipids in the tissues caused an increase in the levels of antioxidant enzyme activities, thus increasing the germination and seedling emergence performance of pepper seeds [119]. Xu et al. [120] reported that external MEL applications caused a significant increase in enzymatic antioxidants such as SOD, POX, CAT, and APX peroxidase and nonenzymatic antioxidants such as ascorbic acid and vitamin E, resulting in decreased ROS levels and lipid peroxidation in cucumber under high temperature stress. Lei et al. [114] reported that MEL applications improved attenuates cold-induced apoptosis root cell suspensions in a process that does not relate to reactive oxygen species generation in carrot.

Posmyk et al. [121] investigated osmo- and hydropriming with MEL application on germination in cucumber (*Cucumis sativus*) in order to improve germination under cold stress conditions. Seed germination increased to 50–60% at 15°C and the addition of 25–100 μM MEL increased the germination percentage. Following these results, it was reported that MEL treatments protected cell membranes against peroxidation in cucumber seeds during chilling stress but high levels of MEL caused oxidative changes in proteins. The mitigating role of MEL in two bermudagrass (*Cynodon dactylon*) genotypes under lower temperatures was reported. The effects of MEL differed between genotypes, which were attributed to differential adaptive responses to lower temperatures due to differentiation of antioxidant enzyme activity, photosystem capacity, and metabolic homeostasis [122].

4. Conclusion

Based on literature knowledge, MEL, which is considered a plant growth regulator candidate and known as tolerance to stress in plants, can be used to increase the plant productivity positively under the abiotic stress conditions. It enhances plant growth such as shoot and root biomass, induces root formation, and increases seed germination under unfavorable conditions. These positive attributes could be caused by (1) improving photosynthetic capacity, (2) reducing oxidative stress, (3) enhancing antioxidant activity, (4) downregulating or upregulating stress-related genes, and (4) elevating osmotic metabolites. There are still many unanswered questions about MEL and more areas for further research. The mechanisms by which MEL is produced are still largely unresolved and need to be elucidated by different plant cells in different situations.

Author details

Raziye Kul[1], Aslıhan Esringü[2], Esin Dadasoglu[3], Üstün Sahin[4], Metin Turan[5], Selda Örs[4], Melek Ekinci[1], Guleray Agar[6] and Ertan Yildirim[1*]

1 Faculty of Agriculture, Department of Horticulture, Atatürk University, Erzurum, Turkey

2 Architecture and Design Faculty, Department of Landscape Architecture, Atatürk University, Erzurum, Turkey

3 Faculty of Agriculture, Department of Crop Science, Atatürk University, Erzurum, Turkey

4 Faculty of Agriculture, Department of Agricultural Structures and Irrigation, Atatürk University, Erzurum, Turkey

5 Faculty of Engineering and Architecture, Department of Genetics and Bioengineering, Yeditepe University, Istanbul, Turkey

6 Faculty of Science, Department of Biology, Atatürk University, Erzurum, Turkey

*Address all correspondence to: ertanyil@atauni.edu.tr

References

[1] Gürel A, Avcıoğlu R. Bitkilerde strese dayanıklılık fizyolojisi. In: Özcan S, Gürel E, Babaoğlu M, editors. Bitki Biyoteknolojisi II, Genetik Mühendisliği ve Uygulamaları, 21. Bölüm. Konya: Selçuk University Foundation; 2001. pp. 308-313

[2] Denby K, Gehring C. Engineering drought and salinity tolerance in plants: Lessons from genome-wide expression profiling in Arabidopsis. Trends in Biotechnology. 2005;**23**(11):547-552. DOI: 10.1016/j.tibtech.2005.09.001

[3] Kalefetoğlu T, Ekmekçi Y. The effects on drought on plants and tolerance mechanisms. Gazi University Journal of Science. 2005;**18**:723-740

[4] Boyer JS. Plant productivity and environment potential for increasing crop plant productivity, genotypic selection. Science. 1982;**218**:443-448

[5] Mahajan S, Tuteja N. Cold, salinity ve drought stres: An overwiev. Archives of Biochemistry and Biophysics. 2005;**444**:139-158. DOI: 10.1016/j.abb.2005.10.018

[6] Lerner AB, Case JD, Takahashi Y, Lee TH, Mori W. Isolation of melatonin, a pineal factor that lightens melanocytes. Journal of the American Chemical Society. 1958;**80**:2587

[7] Dubbels R, Reiter RJ, Klenke E, Goebel A, Schnakenberg E, Ehlers C. Melatonin in edible plants identified by radioimmunoassay and by high performance liquid chromatography mass spectrometry. Journal of Pineal Research. 1995;**18**:28-31. DOI: 10.1111/j.1600-079X.1995.tb00136.x

[8] Hattori A, Migitaka H, Masayaki I, Itoh M, Yamamoto K, Ohtani-Kaneko R. Identification of melatonin in plant seed its effects on plasma melatonin levels and binding to melatonin receptors in vertebrates. Biochemistry and Molecular Biology International. 1995;**35**:627-634

[9] Erland LA, Murch SJ, Reiter RJ, Saxena PK. A new balancing act: The many roles of melatonin and serotonin in plant growth and development. Plant Signaling & Behavior. 2015;**10**:e1096469. DOI: 10.1080/15592324.2015.1096469

[10] Reiter R, Tan D-X, Zhou Z, Cruz M, Fuentes-Broto L, Galano A. Phytomelatonin: Assisting plants to survive and thrive. Molecules. 2015;**20**:7396-7437. DOI: 10.3390/molecules20047396

[11] Hardeland R. Melatonin in plants—Diversity of levels and multiplicity of functions. Frontiers in Plant Science. 2016;7:198. DOI: 10.3389/fpls.2016.00198

[12] Arnao MB, Hernandez-Ruiz J. Melatonin: Plant growth regulator and/or biostimulator during stress? Trends in Plant Science. 2014;**19**:789-797. DOI: 10.1016/j.tplants.2014.07.006

[13] Nawaz MA, Huang Y, Bie Z, Ahmed W, Reiter RJ, Niu M. Melatonin: Current status and future perspectives in plant science. Frontiers in Plant Science. 2016;**6**:1230. DOI: 10.3389/fpls.2015.01230

[14] Murch SJ, KrishnaRaj S, Saxena PK. Tryptophan is a precursor for melatonin and serotonin biosynthesis in in vitro regenerated St. John's Wort (*Hypericum perforatum* L. Cv. Anthos) plants. Plant Cell Reports. 2000;**19**:698-704

[15] Paredes SD, Marchena AM, Bejarano I, Espino J, Barriga C, Rial RV, et al. Melatonin and tryptophan affect the activity-rest rhythm, core and peripheral temperatures, and interleukin levels

in the ringdove: Changes with age. The Journals of Gerontology. Series A, Biological Sciences and Medical Sciences. 2009;**63**:340-350. DOI: 10.1093/gerona/gln054

[16] Murch SJ, Simmons CB, Saxena PK. Melatonin in feverfew and other medicinal plants. Lancet. 1997;**350**:1598-1599

[17] Antoniou C, Chatzimichail G, Xenofontos R, Pavlou G, Panagiotou E, Christou A, et al. Melatonin systemically ameliorates drought stress-induced damage in *Medicago sativa* plants by modulating nitro-oxidative homeostasis and proline metabolism. Journal of Pineal Research. 2017;**62**:e12401. DOI: 10.1111/jpi.12401

[18] Wang P, Sun X, Li C, Wei Z, Liang D, Ma F. Long-term exogenous application of melatonin delays drought-induced leaf senescence in apple. Journal of Pineal Research. 2013;**54**:292-302. DOI: 10.1111/jpi.12017

[19] Wei W, Li Q-T, Chu Y-N, Reiter RJ, Yu X-M, Zhu D-H, et al. Melatonin enhances plant growth and abiotic stress tolerance in soybean plants. Journal of Experimental Botany. 2014;**66**:695-707. DOI: 10.1093/jxb/eru392

[20] Arnao MB, Hernández-Ruiz J. Chemical stress by different agents affects the melatonin content of barley roots. Journal of Pineal Research. 2009;**46**:295-299. DOI: 10.1111/j.1600-079X.2008.00660.x

[21] Chen Q, Qi W, Reiter RJ, Wei W, Wang B. Exogenously applied melatonin stimulates rootgrowth and raises endogenous indoleacetic acid inroots of etiolated seedlings of *Brassica juncea*. Journal of Plant Physiology. 2009;**166**:324-328. DOI: 10.1016/j.jplph.2008.06

[22] Park S, Back K. Melatonin promotes seminal root elongation and root growth in transgenic rice after germination. Journal of Pineal Research. 2012;**53**:385-389. DOI: 10.1111/j.1600-079X.2012.01008.x

[23] Liang C, Zheng G, Li W, Wang Y, Hu B, Wang H, et al. Melatonin delays leaf senescence and enhances salt stress tolerance in rice. Journal of Pineal Research. 2015;**59**:91-101. DOI: 10.1111/jpi.12243

[24] Zhang J, Zeng B, Mao Y, Kong X, Wang X, Yang Y, et al. Melatonin alleviates aluminum toxicity through modulating antioxidative enzymes and enhancing organic acid anion exudation in soybean. Functional Plant Biology. 2017;**44**:961-968. DOI: 10.1071/FP17003

[25] Fleta-Soriano E, Diaz L, Bonet E, Munne-Bosch S. Melatonin may exert a protective role against drought stress in maize. Journal of Agronomy and Crop Science. 2017;**203**:286-294. DOI: 10.1111/jac.12201

[26] Hernandez-Ruiz J, Cano A, Arnao MB. Melatonin acts as a growth-stimulating compound in some monocot species. Journal of Pineal Research. 2005;**39**:137-142. DOI: 10.1111/j.1600-079X.2005.00226.x

[27] Liang C, Li A, Yu H, Li W, Liang C, Guo S, et al. Melatonin regulates root architecture by modulating auxin response in rice. Frontiers in Plant Science. 2017;**8**:134. DOI: 10.3389/fpls.2017.00134

[28] Li H, Chang J, Chen H, Wang Z, Gu X, Wei C, et al. Exogenous melatonin confers salt stress tolerance to watermelon by improving photosynthesis and redox homeostasis. Frontiers in Plant Science. 2018;**8**:295. DOI: 10.3389/fpls.2017.00295

[29] Arnao MB, Hernandez-Ruiz J. Melatonin in plants. Plant Signaling & Behavior. 2007;**2**(5):381-382

[30] Murch SJ, Campbell SS, Saxena PK. The role of serotonin and melatonin

in plant morphogenesis: Regulation of auxininduced root organogenesis in in vitro-cultured explants of St. John's Wort (*Hypericum perforatum* L.). In Vitro Cellular & Developmental Biology. Plant. 2001;**37**:786-793

[31] Tan D-X, Hardeland R, Manchester LC, Korkmaz A, Ma S, Rosales-Corral S, et al. Functional roles of melatonin in plants, and perspectives in nutritional and agricultural science. Journal of Experimental Botany. 2012;**63**:577-597. DOI: 10.1093/jxb/err256

[32] Arnao MB, Hernández-Ruiz J. The physiological function of melatonin in plants. Plant Signaling & Behavior. 2006;**1**:89-95. DOI: 10.4161/psb.1.3.2640

[33] Tan DX, Manchester LC, Helton P, Reiter RJ. Phytoremediative capacity of plants enriched with melatonin. Plant Signaling & Behavior. 2007;**2**:514-516

[34] Shi H, Tan DX, Reiter RJ, Ye T, Yang F, Chan Z. Melatonin induces class A1 heat shock factors (HSFA1s) and their possible involvement of thermotolerance in Arabidopsis. Journal of Pineal Research. 2015;**58**:335-342. DOI: 10.1111/jpi.12219

[35] Leyva R, Sánchez-Rodríguez E, Ríos JJ, Rubio-Wilhelmi MM, Romero L, Ruiz JM, et al. Beneficial effects of exogenous iodine in lettuce plants subjected to salinity stress. Plant Science. 2011;**181**:195-202. DOI: 10.1016/j.plantsci.2011.05.007

[36] Munns R, Tester M. Mechanisms of salinity tolerance. Annual Review of Plant Biology. 2008;**59**:651-681. DOI: 10.1146/annurev.arplant.59.032607.092911

[37] Flowers T, Galal H, Bromham L. Evolution of halophytes: Multiple origins of salt tolerance in land plants. Functional Plant Biology. 2010;**37**:604-612

[38] Yildirim E, Taylor AG, Spittler TD. Ameliorative effects of biological treatments on growth of squash plants under salt stress. Scientia Horticulturae. 2006;**111**:1-6. DOI: 10.1016/j.scienta.2006.08.003

[39] Cirillo C, Rouphael Y, Caputo R, Raimondi G, Sifola MI, De Pascale S. Effects of high salinity and the exogenous application of an osmolyte on growth, photosynthesis, and mineral composition in two ornamental shrubs. The Journal of Horticultural Science and Biotechnology. 2016;**91**:14-22. DOI: 10.1080/14620316.2015.1110988

[40] Parida AK, Das AB. Salt tolerance and salinity effects on plants: A review. Ecotoxicology and Environmental Safety. 2005;**60**:324-349. DOI: 10.1016/j.ecoenv.2004.06.010

[41] Acosta-Motos JR, Ortuño MF, Bernal-Vicente A, Diaz-Vivancos P, Sanchez-Blanco MJ, Hernandez JA. Plant responses to salt stress: Adaptive mechanisms. Agronomy. 2017;**7**:18. DOI: 10.3390/agronomy7010018

[42] Gao S, Ouyang C, Wang S, Xu Y, Tang L, Chen F. Effects of salt stress on growth, antioxidant enzyme and phenylalanine ammonia-lyase activities in *Jatropha curcas* L seedlings. Plant, Soil and Environment. 2008;**54**:374-381

[43] Arora N, Bhardwaj R, Sharma P, Arora HK. Effects of 28-homobrassinolide on growth, lipid peroxidation and antioxidative enzyme activities in seedlings of *Zea mays* L. under salinity stress. Acta Physiologiae Plantarum. 2008;**30**:833-839

[44] Yu Y, Wang A, Li X, Kou M, Wang W, Chen X, et al. Melatonin-stimulated triacylglycerol breakdown and energy turnover under salinity stress contributes to the maintenance of plasma membrane H^+–ATPase activity and K^+/Na^+ homeostasis in sweet potato. Frontiers in Plant Science. 2018;**9**:256

[45] Kanwar MK, Yu J, Zhou J. Phytomelatonin: Recent advances and future prospects. J Pineal Res. 2018;**65**:e12526. Available from: https://doi.org/10.1111/jpi.12526

[46] Wang LY, Liu JL, Wang WX, Sun Y. Exogenous melatonin improves growth and photosynthetic capacity of cucumber under salinity-induced stress. Photosynthetica. 2016;**54**:19-27. DOI: 10.1007/s11099-015-0140-3

[47] Dawood MG, EL-Awadi ME. Alleviation of salinity stress on *Vicia faba* L. plants via seed priming with melatonin. Acta Biológica Colombiana. 2015;**20**(2):223-235. DOI: 10.15446/abc.v20n2.43291

[48] Zhou X, Zhao H, Cao K, Hu L, Du T, Baluška F, et al. Beneficial roles of melatonin on redox regulation of photosynthetic electron transport and synthesis of D1 protein in tomato seedlings under salt stress. Frontiers in Plant Science. 2016;**7**:1-10. DOI: 10.3389/fpls.2016.01823

[49] Zhang R, Sun Y, Liu Z, Jin W, Sun Y. Effects of melatonin on seedling growth, mineral nutrition, and nitrogen metabolism in cucumber under nitrate stress. Journal of Pineal Research. 2017;**62**:e12403. DOI: 10.1111/jpi.12403

[50] Ke Q, Ye J, Wang B, Ren J, Yin L, Deng X, et al. Melatonin mitigates salt stress in wheat seedlings by modulating polyamine metabolism. Frontiers in Plant Science. 2018;**9**:1-11. DOI: 10.3389/fpls.2018.00914

[51] Li X, Yu B, Cui Y, Yin Y. Melatonin application confers enhanced salt tolerance by regulating Na^+ and Cl^- accumulation in rice. Plant Growth Regulation. 2017;**83**:441-454. DOI: 10.1007/s10725-017-0310-3

[52] Jiang J, Cui Q, Feng K, Xu D, Li C, Zheng Q. Melatonin improves antioxidant capacity and ion homeostasis and enhances salt tolerance in maize seedlings. Acta Physiologiae Plantarum. 2016;**38**:82. 1-9. DOI: 10.1007/s11738-016-2101-2

[53] Zhang Y, Shen Y. Wastewater irrigation: Past, present, and future. Wiley Interdisciplinary Reviews: Water. 2017:e1234. DOI: 10.1002/wat2.1234

[54] FAO. Properties and Management of Dry Lands. Rome, Italy: Food and Agriculture Organization; 2005

[55] Kijne JW. Abiotic stress and water scarcity: Identifying and resolving conflicts from plant level to global level. Field Crops Research. 2006;**97**:3-18

[56] Türkeş M. Artan sera etkisinin Türkiye üzerindeki etkileri. Tübitak Bilim ve Teknik Dergisi. 1994;**321**:71

[57] Farooq M, Wahid A, Kobayashi N, Fujita D, Basra SMA. Plant drought stress: Effects, mechanisms and management. Agronomy for Sustainable Development. 2009;**29**:185-212

[58] Bajaj S, Jayaprakash T, Li L, Ho TH, Wu R. Transgenic approaches to increase dehydration-stress tolerance in plants. Molecular Breeding. 1999;**5**:493-503

[59] Arora A, Sairam RK, Srivastava GC. Oxidative stress and antioxidative systems in plants. Current Science. 2002;**82**:1227-1238

[60] Escuredo IP, Arrese-Igor C, Becana M. Oxidative damage in pea plants exposed to water deficit or paraquat. Plant Physiology. 1998;**116**:173-181

[61] Monakhova OF, Chernyadev II. Protective role of kartolin-4 in wheat plants exposed to soil drought. Applied and Environmental Microbiology. 2002;**38**:373-380

[62] Li H, He J, Yang X, Li X, Luo D, Wei C. Glutathione-dependent

induction of local and systemic defense against oxidative stress by exogenous melatonin in cucumber (Cucumis sativus L.). Journal of Pineal Research. 2016;**60**:206-216. DOI: 10.1111/jpi.12304

[63] Li J, Zeng L, Cheng Y, Lu G, Fu G, Ma H, et al. Exogenous melatonin alleviates damage from drought stress in *Brassica napus* L. (rapeseed) seedlings. Acta Physiologiae Plantarum. 2018;**40**:43. DOI: 10.1007/s11738-017-2601-8

[64] Cui G, Zhao X, Liu S, Sun F, Zhang C, Xi Y. Beneficial effects of melatonin in overcoming drought stress in wheat seedlings. Plant Physiology and Biochemistry. 2017;**118**:138-149. DOI: 10.1016/j.plaphy.2017.06.014

[65] Meng JF, Xu TF, Wang ZZ, Fang YL, Xi ZM, Zhang ZW. The ameliorative effects of exogenous melatonin on grape cuttings under water-deficient stress: Antioxidant metabolites, leaf anatomy, and chloroplast morphology. Journal of Pineal Research. 2014;**57**:200-212. DOI: 10.1111/jpi.12159

[66] Ma X, Zhang J, Burgess P, Rossi S, Huang B. Interactive effects of melatonin and cytokinin on alleviating drought induced leaf senescence in creeping bentgrass (*Agrostis stolonifera*). Environmental and Experimental Botany. 2018;**145**:1-11. DOI: 10.1016/j.envexpbot.2017.10.010

[67] Li C, Tan DX, Liang D, Chang C, Jia D, Ma F. Melatonin mediates the regulation of ABA metabolism, free-radical scavenging, and stomatal behavior in two *Malus* species under drought stress. Journal of Experimental Botany. 2015;**66**:669-680

[68] Li D, Zhang D, Wang H, Li Y, Li R. Physiological response of plants to polyethylene glycol (PEG-6000) by exogenous melatonin application in wheat. Zemdirbyste-Agriculture. 2017;**104**(3):219-228. DOI: 10.13080/z-a.2017.104.028

[69] Seki M, Umezawa T, Urano K, Shinozaki K. Regulatory metabolic networks in drought stress responses. Current Opinion in Plant Biology. 2007;**10**:296-302

[70] Wang L, Zhao Y, Reiter RJ, He C, Liu G, Lei Q. Changes in melatonin levels in transgenic 'Micro-Tom' tomato overexpressing ovine AANAT and ovine HIOMT genes. Journal of Pineal Research. 2014;**56**:134-142. DOI: 10.1111/jpi.12105

[71] Rubio MI, Escrig I, Martínez-Cortina C, López-Benet FJ, Sanz A. Cadmium and nickel accumulation in rice plants. Effects on mineral nutrition and possible interactions of abscisic and gibberellic acids. Plant Growth Regulation. 1994;**14**(2):151-157

[72] Prasad MNV. Cadmium toxicity and tolerance in vascular plants. Environmental and Experimental Botany. 1995;**35**:525-545

[73] Salt D. Responses and adaptations of plants to metal stress. In: Hawkesford MJ, editor. Molecular Analysis of Plant Adaptations to the Environment. Dordrecht: Kluwer Academic Publishers; 2001. pp. 159-179

[74] Prasad MNV, Strazalka K. Physiology and Biochemistry of Metal Toxicity and Tolerance in Plants. Dordrecht: Kluwer Academic Publishers; 2002. 432 p. ISBN 1-40-200468-0

[75] Brune A, Urbach W, Dietz KJ. Differential toxicity of heavy metals is partly related to a loss of preferential extraplasmic compartmentation: A comparison of Cd-, Mo-, Ni-, and Zn-stress. New Phytologist. 1995;**129**:404-409

[76] Buchanan BB, Gruissen W, Jones RL. Biochemistry and Molecular Biology of Plants. Rockville: American Society of Plant Physiology; 2000. pp. 1-367

[77] Loureiro S, Santos C, Pinto G, Costa A, Monteiro M, Nogueira AJA, et al. Toxicity assessment of two soils from Jales mine (Portugal) using plants: Growth and biochemical parameters. Archives of Environmental Contamination and Toxicology. 2006;**50**:182-190

[78] Hu K, Hu LY, Li YH, Zhang FQ, Zhang H. Protective roles of nitric oxide on germination and antioxidant metabolism in wheat seeds under copper stress. Plant Growth Regulation. 2007;**53**:173-183

[79] Gupta M, Sharma P, Sarin NB, Sinha AK. Differential response of arsenic stress in two varieties of *Brassica junce* a L. Chemosphere. 2009;**74**:1201-1208

[80] Tal O, Haim A, Harel O, Gerchman Y. Melatonin as an antioxidant and its semi-lunar rhythm in green macroalga Ulva sp. Journal of Experimental Botany. 2011;**62**:1903-1910. DOI: 10.1093/jxb/erq378

[81] Posmyk MM, Kuran H, Marciniak K, Janas KM. Presowing seed treatment with melatonin protects red cabbage seedlings against toxic copper ion concentrations. Journal of Pineal Research. 2008;**45**:24-31

[82] Nawaz MA, Jiao Y, Chen C, Shireen F, Zheng Z, Imtiaz M, et al. Melatonin pretreatment improves vanadium stress tolerance of watermelon seedlings by reducing vanadium concentration in the leaves and regulating melatonin biosynthesis and antioxidant-related gene expression. Journal of Plant Physiology. 2018;**220**:115-127. DOI: 10.1016/j.jplph.2017.11.003

[83] Tang Y, Li J, Li H. Effects of exogenous melatonin on photosynthetic characteristics of eggplant (*Solanum melongena* L.) under cadmium stress. In: International Conference on Manufacturing Science and Engineering (ICMSE 2015); 2015

[84] Hasan MK, Ahammed GJ, Yin L, Shi K, Xia X, Zhou Y, et al. Melatonin mitigates cadmium phytotoxicity through modulation of phytochelatins biosynthesis, vacuolar sequestration, and antioxidant potential in *Solanum lycopersicum* L. Frontiers in Plant Science. 2015;**6**:601. DOI: 10.3389/fpls.2015.00601

[85] Gu Q, Chen Z, Yu X, Cui W, Pan J, Zhao G, et al. Melatonin confers plant tolerance against cadmium stress via the decrease of cadmium accumulation and reestablishment of microRNA-mediated redox homeostasis. Plant Science. 2017;**261**:28-37. DOI: 10.1016/j.plantsci.2017.05.001

[86] Sarafi E, Tsouvaltzis P, Chatzissavvidis C, Siomos A, Therios I. Melatonin and resveratrol reverse the toxic effect of high boron (B) and modulate biochemical parameters in pepper plants (*Capsicum annuum* L.). Plant Physiology and Biochemistry. 2017;**112**:173-182. DOI: 10.1016/j.plaphy.2016.12.018

[87] Ni J, Wang Q, Shah FA, Liu W, Wang D, Huang S, et al. Exogenous melatonin confers cadmium tolerance by counterbalancing the hydrogen peroxide homeostasis in wheat seedlings. Molecules. 2018;**23**:799. DOI: 10.3390/molecules23040799

[88] Moussaa HR, Algamal SMA. Does exogenous application of melatonin ameliorate boron toxicity in spinach plants? International Journal of Vegetable Science. 2017;**23**(3):233-245. DOI: 10.1080/19315260.2016.1243184

[89] Pierce LC. Vegetables. Characteristics, Production and Marketing. USA: John Willey and Sons Inc.; 1987. 433 p

[90] Decoteau DR. Vegetable Crops. New Jersey, USA: Prentice-Hall Inc.; 2000

[91] Kratsch HA, Wise RR. The ultrastructure of chilling stress. Plant, Cell & Environment. 2000;**23**:337-350

[92] Lyons JM. Chilling injury in plants. Annual Review of Plant Physiology. 1973;**24**:445-466

[93] Esim N, Atici O. Nitric oxide improves chilling tolerance of maize by affecting apoplastic antioxidative enzymes in leaves. Plant Growth Regulation. 2014;**72**:29-38. DOI: 10.1007/s10725-013-9833-4

[94] Kusvuran S, Ellialtioglu S, Polat Z. Antioxidative enzyme activity, lipid peroxidation, and proline accumulation in the callus tissues of salt and drought tolerant and sensitive pumpkin genotypes under chilling stress. Horticulture, Environment and Biotechnology. 2013;**54**:319-325. DOI: 10.1007/s13580-013-1042-6

[95] Liu Y, Jiang H, Zhao Z, An L. Nitric oxide synthase like activity-dependent nitric oxide production protects against chilling induced oxidative damage in *Chorispora bungeana* suspension cultured cells. Plant Physiology and Biochemistry. 2010;**48**:936-944. DOI: 10.1016/j.plaphy.2010.09.001

[96] Fan J, Ren J, Zhu W, Amombo E, Fu J, Chen L. Antioxidant responses and gene expression in bermudagrass under cold stress. Journal of the American Society for Horticultural Science. 2014;**139**:699-705

[97] Almeselmani M, Deshmukh PS, Sairam RK, Kushwaha SR, Singh TP. Protective role of antioxidant enzymes under high temperature stress. Plant Science. 2006;**171**:382-388

[98] Yin H, Chen Q, 2008 YM. Effects of short-term heat stress on oxidative damage and responses of antioxidant system in Lilium longiflorum. Plant Growth Regulation;**54**:45-54

[99] Al-Khatib K, Paulsen GM. High-temperature effects on photosynthetic processes in temperate and tropical cereals. Crop Science. 1999;**39**:119-125

[100] Sam O, Nunez M, Ruiz-Sancchez MC, DellAmico J, Falcon V, DeLaRosa MC, et al. Effect of a brassinosteroid analogue and high temperature stres on leaf ultrastructure of *Lycopersicon esculentu*m. Biologia Plantrum. 2000;**44**(2):213-218

[101] Hall AE. Crop Responses to Environment. Boca Raton, Florida: CRC Press LLC; 2001

[102] Rosa M, Rivero ES, Ruiz JM, Romero L. Influence of temperature on biomass, iron metabolism and some related bioindicators in tomato and watermelon plants. Journal of Plant Physiology. 2003;**160**:1065-1071

[103] Hall AE. Breeding for heat tolerance. Plant Breeding Reviews. 1992;**10**:129-168

[104] Hall AE. Physiology and breeding for heat tolerance in cowpea, and comparison with other crops. In: Kuo CG, editor. Adaptation of Food Crops to Temperature and Water Stress. Shanhua, Taiwan: Asian Vegetable Research and Development Center; 1993. pp. 271-284. Publ. No. 93-410

[105] Suzuki N, Mittler R. Reactive oxygen species and temperature stresses: A delicate balance between signaling and destruction. Physiologia Plantarum. 2006;**126**:45-51

[106] Tan DX, Manchester LC, Reiter RJ, Qi WB, Karbownik M, Calvo JR. Significance of melatonin in antioxidative defense system: Reactions and products. Biological Signals and Receptors. 2000;**9**:137-159

[107] Arnao MB, Hernandez-Ruiz J. Melatonin and its relationship to plant hormones. Annals of Botany. 2018;**121**:195-207. DOI: 10.1093/aob/mcx114

[108] Janas K, Posmyk M. Melatonin, an underestimated natural substance

with great potential for agricultural application. Acta Physiologiae Plantarum. 2013;**35**:3285-3292

[109] Bajwa VS, Shukla MR, Sherif SM, Murch SJ, Saxena P. Role of melatonin in alleviating cold stress in Arabidopsis thaliana. Journal of Pineal Research. 2014;**56**:238-245. DOI: 10.1111/jpi.12115

[110] Shi H, Chan Z. The cysteine2/histidine2-type transcription factor zinc finger of Arabidopsis thaliana 6-activated C-repeat-binding factor pathway is essential for melatonin-mediated freezing stress resistance in Arabidopsis. Journal of Pineal Research. 2014;**57**:185-191. DOI: 10.1111/jpi.12155

[111] Turk H, Erdal S, Genisel M, Atici O, Demir Y, Yanmis D. The regulatory effect of melatonin on physiological, biochemical and molecular parameters in cold-stressed wheat seedlings. Plant Growth Regulation. 2014;**74**:139-152. DOI: 10.1007/s10725-014-9905-0

[112] Xu W, Cai SY, Zhang Y, Wang Y, Ahammed GJ, Xia XJ, et al. Melatonin enhances thermotolerance by promoting cellular protein protection in tomato plants. Journal of Pineal Research. 2016;**61**:457-469. DOI: 10.1111/jpi.12359

[113] Zhang J, Shi Y, Zhang X, Du H, Xu B, Huang B. Melatonin suppression of heat-induced leaf senescence involves changes in abscisic acid and cytokinin biosynthesis and signaling pathways in perennial ryegrass (*Lolium perenne* L.). Environmental and Experimental Botany. 2017;**138**:36-45. DOI: 10.1016/j.envexpbot.2017.02.012

[114] Lei XY, Zhu RY, Zhang GY, Dai YR. Attenuation of cold-induced apoptosis by exogenous melatonin in carrot suspension cells: The possible involvement of polyamines. Journal of Pineal Research. 2004;**36**:126-131

[115] Bałabusta M, Szafranska K, Posmyk MM. Exogenous melatonin improves antioxidant defense in cucumber seeds (*Cucumis sativus* L.) germinated under chilling stress. Frontiers in Plant Science. 2016;7:575. DOI: 10.3389/fpls.2016.00575

[116] Ding F, Wang M, Liu B, Zhang S. Exogenous melatonin mitigates photoinhibition by accelerating non-photochemical quenching in tomato seedlings exposed to moderate light during chilling. Frontiers in Plant Science. 2017;**8**:244. DOI: 10.3389/fpls.2017.00244

[117] Alam MN, Zhang L, Yang L, Islam R, Liu Y, Luo H, et al. Transcriptomic profiling of tall fescue in response to heat stress and improved thermotolerance by melatonin and 24-epibrassinolide. BMC Genomics. 2018;**19**:224. DOI: 10.1186/s12864-018-4588-y

[118] Kołodziejczyk I, Dzitkob K, Szewczyk R, Posmyka MM. Exogenous melatonin improves corn (*Zea mays* L.) embryo proteome in seeds subjected to chilling stress. Journal of Plant Physiology. 2016;**193**:47-56. DOI: 10.1016/j.jplph.2016.01.012

[119] Korkmaz A, Karaca A, Kocacinar F, Cuci Y. The effects of seed treatment with melatonin on germination and emergence performance of pepper seeds under chilling stress. Tarım Bilimleri Dergisi. 2017;**23**:167-176

[120] Xu XD, Sun Y, Sun B, Zhang J, Guo XQ. Effects of exogenous melatonin on active oxygen metabolism of cucumber seedlings under high temperature stress. Ying Yong Sheng Tai Xue Bao. 2010;**21**:1295-1300. DOI: 10.1155/2012/490647

[121] Posmyk MM, Balabusta M, Wieczorek M, Sliwinska E, Jana KM. Melatonin applied to cucumber (*Cucumis sativus* L.) seeds improves germination during chilling stress. Journal of Pineal

Research. 2009;**46**:214-223. DOI: 10.1111/j.1600-079X.2008.00652.x

[122] Hu Z, Fan J, Xie Y, Amombo E, Liu A, Gitau MM, et al. Comparative photosynthetic and metabolic analyses reveal mechanism of improved cold stress tolerance in bermudagrass by exogenous melatonin. Plant Physiology and Biochemistry. 2016;**100**:94e104. DOI: 10.1016/j.plaphy.2016.01.008

Chapter 9

Regulation Effect of Different Water Supply to the Nitrogen and Carbon Metabolism

Szilvia Veres, László Zsombik and Csaba Juhász

Abstract

Drought stress and flood result in the generation and accumulation of active oxygen species, the peroxidation of membrane lipids, and reduction of nitrogen metabolism, photosynthesis, growth, and development, causing a significant decline in the qualitative and quantitative production. The water availability influences the different component of NUE and photosynthetic system and its connections. The goal of this chapter is to summarize the effect of water supply to the nitrogen and carbon metabolisms. Knowing about the value of nitrogen use efficiency and photosynthetic parameters is really a useful essential for selecting and growing the best genotypes. But what will happen with these two crucial characteristics of plants, if the environment for growing is not ideal?.

Keywords: nitrogen, carbon, drought, water

1. Introduction

The most widely discussed projections—*World Population Prospects* every 2 years—are those published by the United Nations from 1951. From this year they have published knowledge about the world population that increased more than 400% over the twentieth century [1]. Expanding populations, income growth, and urbanization have brought about quantitative and qualitative changes in the demand for food. Agriculture faces multiple challenges: it has to produce more food to feed a growing population and more feedstocks for a potentially huge bioenergy market, adopt more efficient and sustainable production methods, and adapt to climate change. Although all of the challenges are more or less hang together, thus if we will be able to find a good point to step in, it can mean a medicine for all function. Climatic change is the middle of this complex problem; it is the reason and the solution as well. According to the most recent assessment report of the Intergovernmental Panel on Climate Change (IPCC), published in 2014, levels of anthropogenic emissions of greenhouse gases are now at their highest in history [2]. Agricultural production and its effect on land use are major sources of these emissions, by sharing methane and nitrous oxide gases. As more greenhouse gas emissions are released into the air, causing air temperatures to increase, more moisture evaporates from land and water bodies. Warmer temperatures also increase evaporation and evapotranspiration in plants and soils, which affects plant life and can reduce rainfall even more.

 IntechOpen

Plant growth and productivity are adversely affected by water deficiency and/or drought. Moreover, they are not able to solve their serious problem in a short way. Responses of plants to water stress may be assigned as either injurious change or tolerance index. Short-term acclimatization skills are not always enough mainly if any other loads are also presented. Therefore the development of plants with increased survivability and growth during water stress is a major objective in the breeding crops. All of the skills and traits of what the different plant genotypes are able to use need to be known to cope with water deficiency as single or under multiplied stress conditions.

2. Plant response to drought stress

Drought stress is one of the most common abiotic stresses for terrestrial plants [3, 4]. Drought adaptability of plants was defined as comprehensive capacity for adaptation to the drought stress and the re-watering cycle. Drought resistance and recovery determine drought adaptation of plants. Plants can increase the drought resistance through three strategies, namely, "drought escape," "drought avoidance," and "drought tolerance" (**Figure 1**). The "drought escape" strategy plants reduce life span and induce vegetative dormancy to escape severe drought stress [5]. Drought avoidance includes increasing water uptake ability and water use efficiency: stomatal closure, root systems, high capacity for water transport from roots to leaves, and high leaf mass per leaf area [6]. Behind the strategy of drought tolerance, the plant cells improve osmotic adjustment ability, increase cell wall elasticity to maintain tissue turgidity, elevate antioxidant metabolism, and enhance the resistance to xylem cavitation [7].

It is possible to categorize plant responses to drought stress in accordance with the organizational level of study: from whole plant to molecular level. At molecular level: epigenomics, which affect DNA activity without modifying the gene sequence; transcriptomics, which are changes in gene expression; proteomics, referring to changes in proteins; and finally metabolomics, which are changes in metabolites [9]. Epigenetic mechanisms regulate chromatin structure, gene expression, transposon mobility, and DNA recombination [10]. Several authors have reported the differential regulation of genes encoding epigenetic regulators [11–13] as well as local chromatin and DNA methylation changes in response to a variety of abiotic stresses

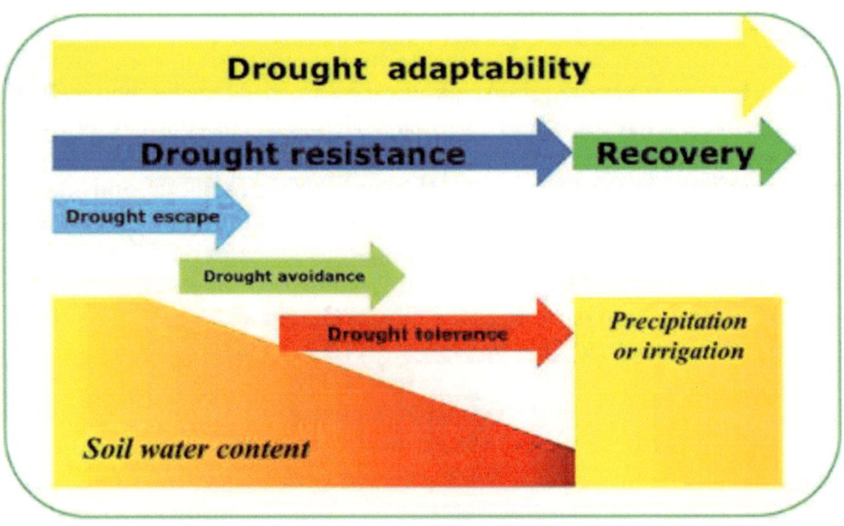

Figure 1.
Diagram about drought-adaptive capabilities during drought and re-watering cycle [8].

[14–16] including drought [17–19]. Recent studies had been performed to investigate drought responses in plants using transcriptomic approaches [20, 21]. The differentially expressed transcripts help for identifying the key genes in stress response and performing functional analysis to reveal their roles in stress adaptation in plants. High-throughput proteomics has proved to be a powerful tool for the comprehensive identification of drought-responsive proteins in plants [22]. In previous investigations, more than 2200 drought-responsive protein species have been identified in leaves [23]; these drought-responsive protein species are defined as 440 unique proteins on the basis of their protein sequence homology and functional domain similarity. Further progress in plant performance under stresses is expected by combining marker-assisted breeding with metabolite markers. The most dramatic effect on the metabolite composition was experienced in leaves compared with other organs [24]. Drought stress evoked the accumulation of many amino acids (glycine, serine) including isoleucine, valine, threonine, and 4-aminobutanoate, which has been reported in both field and greenhouse experiments in many plant species [25].

Plants can adapt to water deficiency by a wide range of alterations in their morphology, anatomy, and physiology that have been the focus of many studies [26, 27]. Plant's strategies on the whole plant level can lead to stomatal closure, reductions in photosynthesis and transpiration, growth inhibition, antioxidant production, and changes in hormonal composition [28–30]. Plants have evolved several strategies to cope with drought stress, including drought escape via a short life cycle or developmental plasticity, drought avoidance via enhanced water uptake and reduced water loss, as well as drought tolerance via osmotic adjustment, antioxidant capacity, and desiccation tolerance [31]. Plant responses to drought stress also vary at different growth stages of the crop [32]. The decrease in yield varies from 13 to 94% in the investigated crops that were under drought stress [33].

3. Interaction between C and N metabolism

Nitrogen is an essential macronutrient for plants, and it can affect many aspects of plant growth and metabolic pathways [34–36]. Nitrate is a primary nitrogen source for photosynthetic organisms. The assimilation of nitrate-N into amino group of amino acids contains at least three main steps: nitrate uptake, reduction of nitrate to ammonium, and incorporation of ammonium to carbon skeleton. One of the main connections is that all assimilatory power utilized by plant metabolism originates from photosynthesis. In plant leaves nitrate assimilation is a direct photosynthetic process that increases the capacity of the photosynthetic apparatus for non-cycling electron flow, overcoming the limitation imposed by CO_2 fixation through the Calvin cycle [37]. Nitrogen assimilation has an impact on the efficiency of CO_2 fixation and the distribution of just fixed carbon among metabolite fractions. Under limiting light intensity condition, there is a strong competition for reducing equivalents between CO_2 and nitrate assimilation [38]. When the available nitrogen source is ammonium, there is no reduction in CO_2 fixation under low light intensity; moreover ammonium has a positive effect on it [39]. Photosynthetic reactions are involved in the synthesis, regulation, and maintenance of the enzymes of nitrate assimilation pathway. Photorespiratory nitrogen metabolism is one of the important aspects of the interactions of carbon and nitrogen.

High carbohydrate content of grain (in maize, wheat) versus low nitrogen content indicates the crucial role of photosynthesis in attaining maximum yield. With the increased use of fertilizer N being closely associated with enhanced crop yields, it seems reasonable that both carbon and nitrogen metabolisms should be considered when attempting to identify factors that limit productivity [40].

4. Water supply and nitrogen nutrition

Water and nitrogen are the most limiting factors in agricultural production in most parts of the world, which are able to determine and influence the carbon metabolism. Nitrogen is a vital structural component of proteins, ribulose-1,5-bisphosphate carboxylase/oxygenase (Rubisco), nucleic acids, chlorophylls, as well as some hormones, and nitrogen fertilization is an essential agronomic management practice to enhance the crop productivity [41]. Primary carbon metabolism is dependent on nitrogen assimilation, because much of the nitrogen pool of the plant is invested in the proteins (structural and enzymes) and chlorophyll of the photosynthetic apparatus. Leaf chlorophyll provides the platform for photosynthetic system [42], whereas Rubisco is the key enzyme involved in the process of photosynthesis [43]. A major part of plant nitrogen is stored in the enzymes participating in the photosynthesis especially Rubisco, which is a key source of N recycling [44]. The limitations in the photosynthetic process as a consequence of intensified drought stress not only impose direct drought stress-induced damages to plants but also result in light-induced oxidative stress. The lower efficiency of photosynthetic system under drought stress leads to the imbalance between absorbed light energy and its utilization in the carbon assimilation process, which in turn spares more electrons triggering the production of reactive oxygen species. Conversely, nitrogen assimilation requires a continuous supply of energy and carbon skeletons. Interconnected metabolic processes make the effect of water deficit on plant nitrogen nutrition status difficult to predict. An adequate assessment of the impacts of drought stress under different nitrogen levels on the physiological activities and yield attributes can provide the valuable insights for wheat cultivation under drought stress [45]. Efficient nitrogen nutrition has been reported to have the potential to alleviate the drought stress damages by maintaining metabolic activities even at low tissue water potential [31]. Abid et al. [46] published that higher N nutrition contributed to drought tolerance in wheat by maintaining higher photosynthetic activities and antioxidative defense system during vegetative growth periods. N-fertilized wheat plants responded more rapidly to increasing drought stress by closing stomata and reducing net photosynthesis [7].

The nitrogen form and the levels of nitrogen available affect root water uptake [47–49]. Synergetic transport has been found between nitrate and water uptake in roots. In plants supplied with nitrogen in both ammonium and nitrate forms, the high nitrogen supply also increased root hydraulic conductance in plants [50, 51]. The radial water transport also can be influenced by nitrogen nutrition. The water flow in apoplastic pathway is blocked by apoplastic barriers at the endodermis, and water flow continues through the symplastic pathway. The deposition of lignin and suberin in the endodermis may affect root hydraulic conductance; Ranathunge et al. [52] demonstrated that high ammonium supply increased this deposition and thus decrease root hydraulic conductance. Drought stress may induce the alkalinization of leaf apoplast, in tomato [53] and hop [54], and especially in plants supplied with high nitrate [55].

5. Conclusion

Nitrogen is an essential nutrient for plants, and it can affect dry matter production by influencing photosynthetic process in several direct and indirect ways. Drought is also a crucial abiotic factor in terms of photosynthesis and also means control for the nutrient nutrition of plants. The effect of water deficit on nitrogen nutrition has been the subject of considerable research on several plants [7, 56, 57]. The interconnected metabolic processes make it difficult to predict the effect of water deficit on plant nitrogen nutrition status and the exact regulation point in the carbon metabolism. In the future we need to know more skills and traits what the different plant genotypes are able to use to cope with water deficiency as single or under multiplied stress conditions.

Acknowledgements

The work was financed by the Higher Education Institutional Excellence Programme (20428-3/2018/FEKUTSTRAT) of the Ministry of Human Capacities in Hungary, within the framework of the 4th thematic programme of the University of Debrecen. The authors thank the support of the "Kiváló malomipari paraméterekkel rendelkező adaptív őszi búza vonalak előállítása" (AGR_PIAC_13-1-2013-0002) grant and the Arid Land Research Center of the Faculty of Agricultural and Food Sciences and Environmental Management at University of Debrecen.

Conflict of interest

We have no "conflict of interest" to declare.

Author details

Szilvia Veres[1*], László Zsombik[2] and Csaba Juhász[3]

1 Department of Agriculture Botany, Crop Physiology and Biotechnology, Faculty of Agricultural and Food Sciences and Environmental Management, Institute of Crop Sciences, University of Debrecen, Debrecen, Hungary

2 Institutes for Agricultural Research and Educational Farm, Research Institute of Nyíregyháza, University of Debrecen, Nyíregyháza, Hungary

3 Faculty of Agricultural and Food Sciences and Environmental Management, Institute of Water and Environmental Management, University of Debrecen, Debrecen, Hungary

*Address all correspondence to: szveres@agr.unideb.hu

References

[1] Roser M. Future Population Growth. Published online at OurWorldInData.org. Retrieved from: https://ourworldindata.org/future-population-growth/2017 [Online Resource]

[2] Porter JR, Xie L, Challinor AJ, Cochrane K, Howden SM, Iqbal MM, et al. Food security and food production systems. In: IPCC. 2014. Climate Change 2014: Impacts, Adaptation, and Vulnerability. Part A: Global and Sectoral Aspects. Contribution of Working Group II to the Fifth Assessment Report of the Intergovernmental Panel on Climate Change. Cambridge, UK/New York, USA: Cambridge University Press; 2014. pp. 485-533

[3] Yordanov I, Velikova V, Tsonev T. Plant response to drought stress. Photosynthetica. 2000;**38**(1):171-186

[4] Xiong L, Wang RG, Mao G, Koczan MJ. Identification of drought tolerance determinants by genetic analysis of root response to drought stress and abscisic acid. Plant Physiology. 2006;**142**:1065-1074. DOI: 10.1104/pp.106.084632

[5] Geber MA, Dawson TE. Genetic variation in and covariation between leaf gas exchange, morphology, and development in *Polygonum arenastrum*, an annual plant. Oecologia. 1990;**85**: 153-158. DOI: 10.1007/BF00319396

[6] Jackson RB, Sperry JS, Dawson TE. Root water uptake and transport: Using physiological processes in global predictions. Trends in Plant Science. 2000;**5**:482-488. DOI: 10.1016/S1360-1385(00)01766-0

[7] Morgan JM. Osmoregulation and water stress in higher plants. Annual Review of Plant Physiology. 1984;**35**:299-319. DOI: 10.1146/annurev.pp.35.060184.001503

[8] Chen D, Wang S, Cao B, Cao D, Leng G, Li H, et al. Genotypic variation in growth and physiological response to drought stress and re-watering reveals the critical role of recovery in drought adaptation in maize seedlings. Frontiers in Plant Science; 2016;**6**:1241. DOI: 10.3389/fpls.2015.01241

[9] Singh S, Parihar P, Singh R, Singh VP, Prasad SM. Heavy metal tolerance in plants: Role of transcriptomics, proteomics, metabolomics, and ionomics. Frontiers in Plant Sciences. 2015;**6**:1143. DOI: 10.3389/fpls.2015.01143

[10] He G, Elling AA, Deng XW. The epigenome and plant development. Annual Review of Plant Biology. 2011;**62**:411-435

[11] Fang H, Liu X, Thorn G, Duan J, Tian L. Biochemical and biophysical research communications expression analysis of histone acetyltransferases in rice under drought stress. Biochemical and Biophysical Research Communication. 2014;**443**:400-405

[12] Li H, Yan S, Zhao L, Tan J, Zhang Q, Gao F. Histone acetylation associated up-regulation of the cell wall related genes is involved in salt stress induced maize root swelling. BMC Plant Biology. 2014;**14**:105. DOI: 10.1186/1471-2229-14-105

[13] Su LC, Deng B, Liu S, Li LM, Hu B, Zhong YT. Isolation and characterization of an osmotic stress and ABA induced histone deacetylase in *Arachis hypogaea*. Frontiers in Plant Science. 2015;**6**:512. DOI: 10.3389/fpls.2015.00512

[14] Chen L, Luo M, Wang Y, Wu K. Involvement of Arabidopsis histone deacetylase HDA6 in ABA and salt stress response. Journal of Experimental Botany. 2010;**61**:3345-3353

[15] Luo M, Liu X, Singh P, Cui Y, Zimmerli L, Wu K. Chromatin modifications and remodeling in plant abiotic stress responses. Biochimica et Biophysica Acta. 2012;**1819**:129-136

[16] Bocchini M, Bartucca ML, Ciancaleoni S, Mimmo T, Cesco S, Pii Y. Iron deficiency in barley plants: Phytosiderophore release, iron translocation, and DNA methylation. Frontiers in Plant Science. 2015;**6**:514. DOI: 10.3389/fpls.2015.00514

[17] González RM, Ricardi MM, Iusem ND. Epigenetic marks in an adaptive water stress-responsive gene in tomato roots under normal and drought conditions. Epigenetics. 2013;**8**:864-872

[18] Kim JM, Sasaki T, Ueda M, Sako K, Seki M. Chromatin changes in response to drought, salinity, heat, and cold stresses in plants. Frontiers in Plant Science. 2015;**6**:114. DOI: 10.3389/fpls.2015.00114

[19] Fortes AM, Gallusci P. Plant stress responses and phenotypic plasticity in the epigenomics era: Perspectives on the grapevine scenario, a model for perennial crop plants. Frontiers in Plant Science. 2017;**8**:82. DOI: 10.3389/fpls.2017.00082

[20] Zhou J, Wang X, Jiao Y, Qin Y, Liu X, He K. Global genome expression analysis of rice in response to drought and high-salinity stresses in shoot, flag leaf, and panicle. Plant Molecular Biology. 2007;**63**:591-608

[21] Gao F, Wang J, Wei S, Li Z, Wang N, Li H, et al. Transcriptomic analysis of drought stress responses in *Ammopiptanthus mongolicus* leaves using the RNA-seq technique. PLoS One. 2015;**10**(4):e0124382

[22] Mohammadi PP, Moieni A, Komatsu S. Comparative proteome analysis of drought-sensitive and drought-tolerant rapeseed roots and their F1 line under drought stress. Amino Acids. 2012;**43**(5):2137-2152. DOI: 10.1007/s00726-012-1299-6

[23] Wang X, Cai X, Xu C, Wang Q, Dai S. Drought-responsive mechanisms in plant leaves revealed by proteomics. International Journal of Molecular Sciences. 2016;**17**(10):1706

[24] Witt S, Galicia L, Lisec J, Cairns J, Tiessen A, Araus JL, et al. Metabolic and phenotypic responses of greenhouse grown maize hybrids to experimentally controlled drought stress. Molecular Plant. 2012;**5**:401-417

[25] Obata T, Witt S, Lisec J, Palacios-Rojas N, Florez-Sarasa I, Yousfi S, et al. Metabolite profiles of maize leaves in drought, heat, and combined stress field trials reveal the relationship between metabolism and grain yield. Plant Physiology. 2015;**169**:2665-2683

[26] Nagy Z, Tuba Z, Zsoldos F, Erdei L. CO_2-exchange and water relation responses of sorghum and maize during water and salt stress. Journal of Plant Physiology. 1995;**145**:539-544

[27] Gratani L, Bombelli A. Leaf anatomy, inclination, and gas exchange relationships in evergreen sclerophyllous and drought semideciduous shrub species. Photosynthetica. 1999;**37**:573-585. DOI: 10.1023/A:1007171525298

[28] Szegletes ZS, Erdei L, Tari I, Cseuz L. Accumulation of osmoprotectants in wheat cultivars of different drought tolerance. Cereal Research Communications. 2000;**28**:403-410. Available from: http://www.jstor.org/stable/23787271

[29] Lawlor DW, Cornic G. Photosynthetic carbon assimilation and associated metabolism in relation to water deficits in higher plants. Plant, Cell and Environment. 2002;**25**:275-294. DOI: 10.1046/j.0016-8025.2001.00814.x

[30] Zhu JK. Salt and drought stress signal transduction in plants. Annual Review of Plant Biology. 2002;**53**:247-273. DOI: 10.1146/annurev.arplant.53.091401.143329

[31] Zhang LX, Li SX, Zhang H, Liang ZS. Nitrogen rates and drought stress effects on production, lipid peroxidation and antioxidative enzyme activities in two maize (*Zea mays L.*) genotypes. Journal of Agronomy and Crop Science. 2007;**193**:387-397. DOI: 10.1111/j.1439-037X.2007.00276.x

[32] Shi J, Yasuor H, Yermiyahu U, Zuo Q , Ben-Gal A. Dynamic responses of wheat to drought and nitrogen stresses during re-watering cycles. Agricultural Water Management. 2014;**146**:163-172. DOI: 10.1016/j.agwat.2014.08.006

[33] Farooq M, Wahid A, Kobayashi N, Fujita D, Basra S. Plant Drought Stress: Effects, Mechanisms and Management Sustainable Agriculture. Berlin: Springer; 2009. pp. 153-188

[34] Guo S, Kaldenhoff R, Uehlein N, Sattelmacher B, Brueck H. Relationship between water and nitrogen uptake in nitrate- and ammonium-supplied *Phaseolus vulgaris L.* plants. Journal of Plant Nutrition and Soil Science. 2007;**170**:73-80. DOI: 10.1002/jpln.200625073

[35] Wang M, Shen QR, Xu GH, Guo SW. New insight into the strategy for nitrogen metabolism in plant cells. International Review of Cell and Molecular Biology. 2014;**310**:1-37. DOI: 10.1016/B978-0-12-800180-6.00001-3

[36] Ding L, Lu Z, Gao L, Guo S, Shen Q. Is nitrogen a key determinant of water transport and photosynthesis in higher plants upon drought stress? Frontier in Plant Science. 2018;**9**:1143. DOI: 10.3389/fpls.2018.01143

[37] Lara C. Interaction betwee nitrogen and carbon assimilation in green cells. In: Srivastava HS, Singh RP, editors. Nitrogen Nutrition in Higher Plants. New Delhi, India: Associated Publishing Co.; 1995. pp. 323-336. ISBN: 978 81 85211 38

[38] Rathnam CKM. Malate and dihydroxyacetone phosphate-dependent nitrate reduction in spinach leaf chloroplast. Plant Physiology. 1978;**62**(2). DOI: 10.1104/pp.62.2.220

[39] Woo KC, Canvin DT. Effect of ammonia on photosynthetic carbon fixation in isolated spinach leaf. Canadian Journal of Botany. 1980;**48**:389. DOI: 10.1007/s11099-010-0051-2

[40] Swank JC, Below FE, Lambert RJ, Hageman RH. Interaction of carbon and nitrogen metabolism in the productivity of mize. Plant Physiology. 1982;**70**:1185-1190. DOI: 10.1104/pp.70.4.1185

[41] Ata-Ul-Karim ST, Liu X, Lu Z, Yuan Z, Zhu Y, Cao W. In-season estimation of rice grain yield using critical nitrogen dilution curve. Field Crops Research. 2016;**195C**:1-8. DOI: 10.1016/j.fcr.2016.04.027

[42] Murchie EH, Pinto M, Horton P. Agriculture and the new challenges for photosynthesis research. The New Phytologist. 2009;**181**:532-552. DOI: 10.1111/j.1469-8137.2008.02705.x

[43] Makino A. Photosynthesis, grain yield, and nitrogen utilization in rice and wheat. Plant Physiology. 2011;**155**:125-129. DOI: 10.1104/pp.110.165076

[44] Gonzalez-Real MM, Baille A. Changes in leaf photosynthetic parameters with leaf position and nitrogen content within a rose plant canopy (*Rosa hybrida*). Plant, Cell and Environment. 2000;**23**:351-363. DOI: 10.1046/j.1365-3040.2000.00559.x

[45] Teixeira EI, George M, Herreman T, Brown H, Fletcher A, Chakwizira E. The

impact of water and nitrogen limitation on maize biomass and resource-use efficiencies for radiation, water and nitrogen. Field Crops Research. 2014;**168**:109-118. DOI: 10.1016/j.fcr.2014.08.002

[46] Abid M, Tian Z, Ata-Ul-Karim ST, Cui Y, Liu Y, Zahoor R, et al. Nitrogen nutrition improves the potential of wheat (*Triticum aestivum L.*) to alleviate the effects of drought stress during vegetative growth periods. Frontier in Plant Science. 2016;**7**:981. DOI: 10.3389/fpls.2016.00981

[47] Gorska A, Ye Q, Holbrook NM, Zwieniecki MA. Nitrate control of root hydraulic properties in plants: Translating local information to whole plant response. Plant Physiology. 2008;**148**:1159-1167. DOI: 10.1104/pp.108.122499

[48] Gorska A, Lazor JW, Zwieniecka AK, Benway C, Zwieniecki MA. The capacity for nitrate regulation of root hydraulic properties correlates with species' nitrate uptake rates. Plant and Soil. 2010;**337**:447-455. DOI: 10.1007/s11104-010-0540-x

[49] Tyerman SD, Wignes JA, Kaiser BN. Root hydraulic and aquaporin responses to N availability. In: Chaumont F, Tyerman S, editors. Plant Aquaporins. Cham: Springer; 2017. pp. 207-236

[50] Ren B, Wang M, Chen Y, Sun G, Li Y, Shen Q. Water absorption is affected by the nitrogen supply to rice plants. Plant and Soil. 2015;**396**:397-410. DOI: 10.1007/s11104-015-2603-5

[51] Li G, Tillard P, Gojon A, Maurel C. Dual regulation of root hydraulic conductivity and plasma membrane aquaporins by plant nitrate accumulation and high-affinity nitrate transporter NRT2. 1. Plant and Cell Physiology. 2016;**57**:733-742. DOI: 10.1093/pcp/pcw022

[52] Ranathunge K, Schreiber L, Bi Y-M, Rothstein SJ. Ammonium-induced architectural and anatomical changes with altered suberin and lignin levels significantly change water and solute permeabilities of rice (*Oryza sativa L.*) roots. Planta. 2016;**243**:231-249. DOI: 10.1007/s00425-015-2406-1

[53] Jia W, Davies WJ. Modification of leaf apoplastic pH in relation to stomatal sensitivity to root-sourced abscisic acid signals. Plant Physiology. 2007;**143**:68-77. DOI: 10.1104/pp.106.089110

[54] Korovetska H, Novák O, Jùza O, Gloser V. Signalling mechanisms involved in the response of two varieties of *Humulus lupulus L.* to soil drying: I. Changes in xylem sap pH and the concentrations of abscisic acid and anions. Plant and Soil. 2014;**380**:375-387. DOI: 10.1007/s11104-014-2101-1

[55] Wilkinson S, Bacon MA, Davies WJ. Nitrate signalling to stomata and growing leaves: Interactions with soil drying, ABA, and xylem sap pH in maize. Journal of Experimental Botany. 2007;**58**:1705-1716. DOI: 10.1093/jxb/erm021

[56] Gonzales-Dugo V, Durand JL, Gastal F. Water deficit and nitrogen nutrition of crops—A review. Agronomy for Sustainable Development, Springer Verlag/EDP Sciences/INRA. 2010;**30**(3):529-544. DOI: 10.1051/agro/2009059

[57] Veres SZ, Petek M, Makleit P, Kiss L, Gáspár S, Frommer D, et al. Physiological detection of water and nitrogen deprivation. Annals, Series on Agriculture, Silviculture and Veterinary Medicine Sciences. 2017;**6**(1):152-158. ISSN 2069-1149

Chapter 10

Chickpea Abiotic Stresses: Combating Drought, Heat and Cold

Peter Kaloki, Viola Devasirvatham and Daniel K.Y. Tan

Abstract

Chickpea is an important legume providing dietary proteins to both humans and animals. It also ameliorates soil nitrogen through biological nitrogen fixation. Drought, heat and cold are important factors among abiotic stresses limiting production in chickpea. Identification, validation and integration of agronomic, physiological and biochemical traits into breeding programs could lead to increased rates of genetic gain and the development of better adapted cultivars to abiotic stress conditions. This chapter illustrates the effects of stresses on chickpea growth and development. It also reviews the various traits and their relationship with grain yield under stress and proposes recommendation for future breeding.

Keywords: abiotic stresses, chickpea, cold, drought, genomic regions, heat, phenotyping, quantitative trait loci (QTL) and traits

1. Introduction

Chickpea is the third most important food legume globally after dry beans and dry peas [1]. It is grown on 12.4 million hectares (mean of 2010–2012, FAOSTAT) producing 11.3 million tonnes at an average yield of 910 kg/ha (mean of 2010–2012, FAOSTAT) [2]. Chickpea is an important legume in many farming systems and provides biological N fixation which benefits the entire farming system.

However, chickpea production is hampered by biotic and abiotic constraints depending on the ecological region. Among abiotic stresses, drought, heat and cold stresses are the most important yield limiting factors [3, 4]; accounting for up to 50% of chickpea production losses for drought, and 15–20% of yield losses for low and high temperatures [5]. This situation is exacerbated by climate change which may cause higher intensity and frequency of droughts, heat waves and cold spills in the arid and semi-arid areas [6] where chickpea is traditionally cultivated.

Improvements in chickpea abiotic stresses responses have been slowed by the complex inheritance of tolerance, the multi-dimensional nature of stress and the fact that various agronomic, physiological and biochemical changes occur in plants that are difficult to predict [3]. Jha et al. [7] suggest that abiotic stresses are limited by quantitative inheritance and large genotype x environment interaction.

It is important to integrate physiological traits into crop improvement; however, plant breeders have largely focused on empirical selection for yield [8, 9]. This may be a result of the challenges associated with measuring physiological traits compared with other easily measured traits such as photosynthetic rate or carbon isotopic discrimination as well as lack of proper multidisciplinary training [10]. Although traditional plant breeding contributed to increased yields [11], largely attributed to environmental adjustments, genotype and their interaction [12], recent rates of yield increase have started to plateau [13]. However, agronomic, physiological and biochemical traits and their correlation with grain yield under stressed and non-stressed conditions are important to select genotypes by screening [14]. Tolerance to abiotic stresses is influenced by polygenes. Under abiotic stress conditions, many factors and their interaction are involved at a particular time which affects crop growth and yield. To define abiotic stresses in a target environment, the above-mentioned should be considered [15]. This chapter explains the background of major abiotic stresses such as drought, heat and cold and breeding strategies to improve tolerance in chickpea along with present molecular breeding approaches.

2. Drought stress

2.1 The concept of physiological breeding and its current status

Physiological breeding relies on the identification of traits which are later used as indirect selection criteria in core breeding or in introgression programs [8]. Fischer [16] proposed two approaches which can be used for trait identification namely, the black box and the ideotype approaches. The black box approach entails evaluation of genotypes in a specified limiting factor (e.g. drought) and measuring trait associations with economic performance. This method aims to identify and characterise the underlying factors that contribute to the target trait. In contrast, the ideotype approach predicts the desired genotype in the target environment and this is used as a blueprint for cultivar development. The target environment should be as homogeneous as possible and the selection process well managed to reduce confounding factors which may affect trait expression [9]. Nevertheless, most target environments are not homogenous and, in these instances, the most probable environment type, based on long-term environmental records, should be targeted [17].

Target physiological traits for drought response should be easy to measure, highly heritable and correlated with yield. They should not limit yield when conditions are conducive nor have negative pleiotropic effects on other traits of economic importance [18]. As such, developing high yielding cultivars which show plasticity is vital since they can maintain yield in water limited environments and at the same time take advantage of excess moisture during wet years [18]. These physiological traits influence yield through additive gene action [19].

There are several traits that can be targeted in physiological breeding which may help increase yield either directly or indirectly. Some examples include selection for optimised plant height, especially in temperate crops to reduce lodging and to increase harvest index under favourable conditions [18]. Genes that govern plant height have contributed to higher yields by ensuring that more assimilates are partitioned to the grains and the reproductive organs instead of the stem [20]. This reduction in plant height was the key driver of the Green Revolution which saw a quadrupling of wheat and rice yields across much of Asia [21]. Selection for flowering time is also very important, especially in water limited environments since it may have a direct effect on yield [22].

Some research organisations including CIMMYT Mexico are practicing physiological breeding but generally the research arena is still in its infancy. There have been several success stories recently in physiological breeding in wheat, soybean, sunflower, sorghum and maize. These examples are summarised by Richards [18] and include increased axial resistance in wheat whereby reduced xylem vessel diameter in the seminal roots increased hydraulic resistance and thus decreased extraction of water from the subsoil, reduced anthesis to silking interval in maize leading to increased yields, increased ability of soybean to fix nitrogen under drought environments, improved transpiration efficiency in wheat and sunflower via selection for low carbon discrimination, stay green trait in sorghum where plants delay senescence and sanction further uptake of soil water and nitrogen (**Figure 1**).

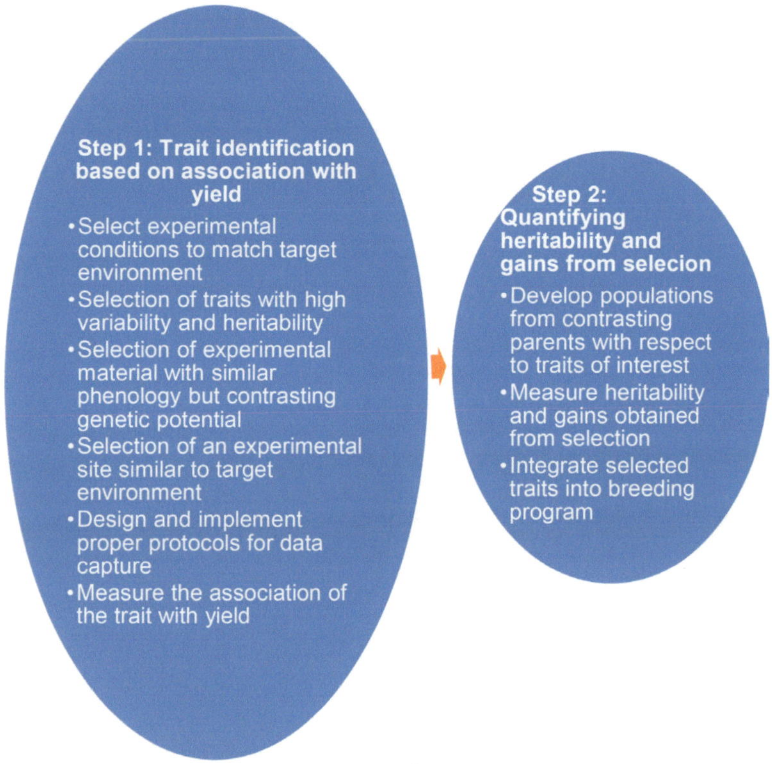

Figure 1.
Schematic illustration of physiological breeding compiled from Reynolds and Trethowan [9].

2.2 Phenotyping target physiological traits in chickpea

Over the last century, breeders have made progress in drought tolerance by selecting constitutive traits that affect dehydration avoidance rather than drought responsive traits because of fewer yield penalties [23]. Target traits in water limited environments should be correlated with yield and should have higher heritability than yield [24]. Phenotyping these traits should also be non-destructive, accurate, cheap and inexpensive [25]. The phenotypic performance needs to be associated with genotypic data to understand the genetic basis of complex these traits [26]. For phenotyping to be successful and relevant, environmental characterisation [25, 27] is vital so that genotype by environment interactions can be exploited [17].

Phenotyping of large plant populations for various traits in the field can be labour intensive and expensive. However, the emergence of high-throughput genotyping platforms such as near infra-red spectroscopy and multi-spectral reflectance makes it possible to phenotype large populations in multi-locations [26].

Chickpea phenotyping for drought tolerance has focused on selection for early maturity to avoid drought and root traits to confer water-use-efficiency under drought [28]. Phenotyping for water use efficiency in chickpea has been conducted using gravimetric methods in a pot culture [28]; however these methods do not generally correlate well with field conditions.

Near infrared spectroscopy has been used to capture differences in dry matter, starch and crude proteins [26]. Spectral reflectance allows monitoring of various dynamic complex traits using high temporal resolution without interfering with the plant [26]. It can be used to measure canopy architecture, nitrogen concentration and water status [26]. Other measurements can be taken on individual plants including plant photosynthesis pigment composition and water status (**Figure 2**) [29].

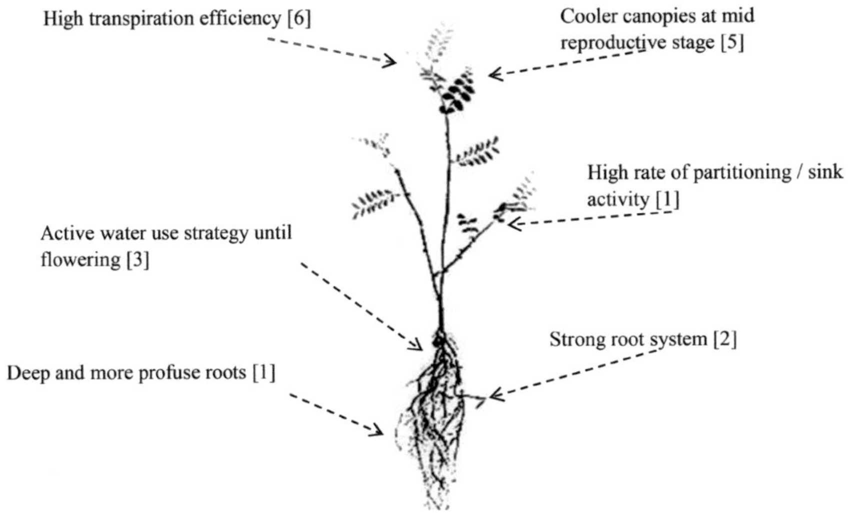

Figure 2.
Some target traits for chickpea physiological breeding.

2.2.1 Canopy temperature

Canopy temperature is an indirect indicator of crop water status since water deficit results in partial stomatal closure, thus reducing transpiration and in effect causing sunlit leaves to become warmer than the ambient temperature [30]. Since transpiration has a cooling effect on canopies, cooler plant canopies indicate higher transpiration rates. This is one of the many factors that affect plant canopy temperature. Others include incident radiation, wind and relative humidity [31]. Under water limited conditions, cooler canopy temperatures are related to the capacity of plants to extract soil water from deep in the subsoil, whereas under well-watered conditions sink strength and photosynthetic capacity are more important [32]. The hand-held canopy temperature gun is a simple and rapid method of determining canopy temperatures. However, in very large populations it may be limiting. Thermal imagery systems are more amenable to high throughput phenotyping for canopy temperature in large populations [33]. Canopy temperature is quite sensitive to environmental conditions and caution should be taken while taking the measurements. Good results are achieved when the conditions are ideal for high vapour pressure deficit (VPD), in conditions of warm air, generally above 15°C and relative humidity of less than 60% with clear sunny skies and low wind speeds [32].

2.2.2 Plant vigour and plant green biomass

Over the years remote sensing imagery has gained popularity because it is not limited by sampling interval or geostatistical interpolation [34], does not involve destructive sampling and it is amenable to high throughput. The premise for using optical remote sensing for crop assessment is that crop canopy multispectral reflectance and temperature is associated with photosynthesis and evaporation whereby leaf area index (LAI) and crop development stage are central [35].

Several indices have been developed which are used to analyse aerial imagery [36] including the Normalised Difference Vegetation Index (NDVI). The NDVI links reflectance in the red region and the NIR to vegetation parameters such as canopy cover, leaf area index and the concentration of total chlorophyll [36]. Korobov and Railyan [37] concluded that the NIR and red areas of the spectrum correlated highly with plant parameters such as plant height, plant density and percent plant cover.

Initially, NDVI was used for estimating green biomass [38]; however, it was subsequently used to assess crop health [39, 40].

2.2.3 Photosynthetically active radiation (PAR)

The photosynthetic active radiation spectrum (PAR), which consists of 50% of total global radiation [41], lies in the wavelength 0.4–0.7 μm [42]. The crop canopy absorbs PAR, referred to as intercepted photosynthetically active radiation (IPAR) to aid photosynthesis, eventually producing biomass [43]. The radiation intercepted during the growing period is determined by the canopy radiation extinction coefficient (k) and is influenced by leaf orientation and the green leaf area [44]. Research has shown that lower k values are associated with narrow and erect leaves compared to plant genotypes with more horizontal leaf arrangements [45]. Lower k values allow more light to penetrate the canopy and illuminate more leaf area in conditions of low light intensity, thus increasing carbon exchange rates and consequently, radiation use efficiency [45].

The fraction of intercepted photosynthetically active radiation can be used to estimate the leaf area index (LAI) by relationship with the plant canopy [43]. This provides an easy and non-destructive way of estimating the leaf area index. IPAR can be accurately determined using a ceptometer, though care should be taken to avoid confounding factors such as the soil albedo, row spacing and lack of canopy uniformity [46].

2.2.4 Chlorophyll content

There is a close relationship between chlorophyll concentration, leaf nitrogen content and crop yield [47]. This relationship arises because the majority of leaf nitrogen is usually contained in the chlorophyll [47]. Since chlorophyll absorbs PAR, which aids in photosynthesis, it indicates the strength of the internal leaf apparatus during photosynthesis [48].

Leaf chlorophyll content can be determined by extraction with organic solvents including acetone [49] and methanol [50] and subsequent quantification using a spectrometer; however this method is expensive and time consuming [51]. A higher throughput non-destructive method is the SPAD chlorophyll meter which allows rapid and inexpensive assessment of leaf greenness [52]. SPAD measures leaf absorbance in the red (650 nm) and infrared (940 nm) regions [53] and gives readings that have been correlated with chlorophyll content under different moisture regimes in many crops [51].

2.2.5 Root traits

Plants extract water from the soil through the roots and the spatial distribution of the root system influences water and nutrient intake capacity [54]. Dense root systems are more efficient at extracting water from the top soil horizon whereas deeper rooting systems better extract water from the lower soil horizons. These contrasting traits are important influences on yield under water deficit conditions during the reproductive stage in many crops [55]. Kashiwagi, Krishnamurthy [56] showed that root architecture affects transpiration by influencing soil moisture use and subsequent harvest index in terminal drought. However, the heritability of these root characteristics will determine their utility in plant breeding. Varshney et al. [57] reported genetic variation for both root length density and root depth in chickpea and found heritabilities ranging from medium to low. Root hydraulic conductivity impacts the amount of water taken up by the plant and is determined by the anatomy and morphology of the roots and their aquaporin activity [58]. In legumes, root hydraulic conductivity is influenced by the total root length since water is absorbed along the full root length [58].

Root phenotyping is difficult and for this reason the literature on chickpea is not extensive. However, Kashiwagi et al. [56] and Zaman-Allah et al. [59] used polyvinyl chloride (PVC) cylinders (lysimeters) to grow chickpeas for assessment. The soil was subsequently washed off from sampled plants to measure total rooting depth. Image analysis software (WinRhizo, Regent Instruments INC., Canada) was then used to estimate the root length at various sections of the lysimeters and divided by the specific volume of that section to determine the root length density.

2.2.6 Transpiration efficiency

Transpiration efficiency is an important yield determinant under drought [60] and as such can be targeted in a breeding program. Carbon isotope discrimination (CID) can be used as a surrogate to measure transpiration efficiency compared with the more tedious gravimetric methods [61].

Carbon isotope composition of plant dry matter was used to evaluate the relative differences of Ci/Ca in C3 species [62]. The stable carbon isotope (^{13}C), which constitutes 1% of total atmospheric CO_2, is usually lower in plant dry matter than in the atmosphere since C3 plants discriminate against ^{13}C during photosynthesis. Carbon isotope discrimination ($\Delta^{13}C$), is the value of the ratio of $^{13}C/^{12}C$ in plant material in relation to the same value of the ratio of the air that plants use [63]. However, transpiration efficiency is a conservative trait that both limits water loss through reduced transpiration and photosynthesis through stomatal closure. While selection for CID will optimise survival under drought stress, yield will be limited under more favourable conditions due to lower stomatal conductance.

2.2.7 Water use efficiency (WUE)

WUE in agriculture can be considered at the whole plant (ratio of total dry matter produced to total water used), economic yield (ratio of crop grain per unit area to transpiration) and leaf (ratio of instantaneous carbon dioxide assimilation rate to transpiration rate at the stomata) levels [64].

At the leaf level, crop water loss is as a result of differences in water vapour concentration between the crop canopy and the atmosphere and is least during cool humid periods of the season. Thus, the rates of CO_2 assimilation (A) and transpiration (T) are a product of stomatal conductance, either of CO_2 (g_c) and water vapour (g_w) or the gradient of either CO_2 ($C_a - C_i$) or water vapour ($W_i - W_a$) between the air outside and inside the leaf [65]. Theoretically, W_T can be improved by lowering the value of C_i/C_a, albeit trade-offs are likely to occur [65]. However, breeders have successfully selected for lower C_i/C_a to produce materials with low stomatal conductance in wheat, higher photosynthetic capacity or a combination of both [62]. There is substantial genetic variation for C_i/C_a determined through $\Delta^{13}C$ which is large enough to cause variation in A/T and consequently WUE for dry matter production [63]. Rebetzke et al. [66] showed that $\Delta^{13}C$ is a highly heritable trait that can be manipulated through plant breeding. Thus increasing intrinsic WUE has been an attractive crop breeding target for many years [16]. By exploiting genetic variation associated with intrinsic earliness and response to photoperiod, breeders have developed genotypes that can grow when the evaporative demand is low, which in turn raises the ratio of A/T and increases yield [67].

Traits associated with water use efficiency have been identified including CID, where low CID implies higher transpiration efficiency resulting from low stomatal conductance, delayed leaf senescence or stay green, spike photosynthesis which results in high water use efficiency due to partial re-fixation of respiratory carbon dioxide, and subcellular processes such as photo-protective mechanisms [68]. These traits can be used as surrogates if yield penalties are not very high.

2.3 Biochemical traits and their relationship with drought tolerance

Production of various secondary metabolites through alteration of plant biochemical tissue profile is one of the ways plants respond to abiotic stresses [69]. The production of the secondary metabolites may be initiated by lipid peroxidation [70] and they help in the protection of membrane lipids from oxidative stress [71]. In order to prevent oxidative damage in plant tissue under water stress conditions, both enzymatic and non-enzymatic antioxidant systems are involved [69]. These include ascorbate, reduced glutathione, β-carotenes, carotenoids, α-tocopherol, peroxidase, catalase, superoxide dismutase, glutathione reductase, polyphenol oxidase and ascorbate peroxidase [69]. The activity levels of the enzymes in the antioxidant system have been used to indicate stress tolerance capabilities for plants exposed to stress conditions [72].

Water stress results in an increase in the production of reactive oxygen species which causes a risk to the plant since it causes the peroxidation of membrane lipids, damage to nucleic acids and the denaturation of proteins [73, 74]. Plants exposed to water stress over a prolonged period of time suffer from extensive cellular damage and eventual death as a result of reactive oxygen species overpowering the scavenging action of the antioxidant system [69].

Hydrogen peroxide plays two roles in plants depending on its concentration. At low levels, it acts as a signal molecule that triggers tolerance to many biotic and abiotic stresses whereas at high levels it causes programmed cell death [69]. Many studies have shown hydrogen peroxide to be an important regulator in many physiological processes including stomatal movement [75], photorespiration and photosynthesis [76], senescence [77], cell division cycle [73] and growth and development [78].

Soluble sugar (fructose, glucose and sucrose) accumulation in plants is closely linked to drought tolerance [69]. Soluble sugars protect plant cells under drought conditions by interacting with proteins and membranes through hydrogen bonding thus inhibiting protein denaturation and secondly, by maintaining structural and functional integrity of macromolecules through vitrification [69].

2.4 Identifying genomic regions linked to physiological traits through association analysis

Genetic association analysis (linkage disequilibrium analysis) is used to identify quantitative trait loci (QTL) by linking the phenotype to the genotype. It exploits historical and evolutionary recombination events at the population level to resolve complex variation up to the sequence level [79]. Phenotypic variation of complex traits in plant species is influenced by multiple QTLs and their interaction with themselves (epistasis) and the environment [80]. Association analysis frequently used [80] because of increased mapping resolution, greater allele number and efficiency [81]. The technique can be applied to a wide array of germplasm from natural populations to study complex trait variation whereas traditional linkage analysis provides information on specific parents [80]. Recent association studies include disease resistance in potato [82], flowering time in maize [83] and iron deficiency chlorosis in soybean [84].

Many QTLs in chickpea for various traits including plant height, plant canopy cover, number of branches per plant, number of pods per plant, 1000 seed weight and days to maturity have been identified using linkage mapping. Gowda et al. [85] identified 41 and 65 QTLs for these traits using single locus QTL analysis from two intraspecific chickpea mapping populations, JG 62 X Vijay and Vijay X ICC 4958, respectively. Rehman et al. [86] found multiple QTLs from a kabuli biparental population of ILC 588 X ILC 3279 for grain yield, harvest index, drought tolerance score, days to flowering, days to maturity and plant height. They also identified two QTLs for stomatal conductance in linkage group (LG) 7 and 3 and six for canopy temperature differential in LG 1, LG 3, LG 4, LG 6 and LG 7. These genomic regions were associated with traits that confer higher productivity under drought stress. The studies of Rehman et al. [86] identified 15 genomic regions associated with drought tolerance traits. Genomic regions on LG 1 and LG 3 were strongly associated with days to flowering, days to maturity, harvest index, reproductive period, canopy temperature differential, plant height and grain yield under drought. Of particular interest was Q3-1 on LG 3 which explained most of the important drought traits. These traits included enhanced stomatal conductance (Q3-3) in LG 3 and cooler canopies (Q1-1 and Q3-2). Hamwieh et al. [87] reported 93 significant QTLs in an intraspecific mapping population developed between ILC 588 and ILC 3279. However, only 8 of these QTLs were expressed in more than one environment. Nevertheless, these authors found a significant QTL region measuring 12.1 cM for days to flowering, drought resistance score, grain yield, days to maturity, harvest index, number of seeds per plant, biological yield and number of pods per plant on LG 3. A second QTL spanning 0.68 cM was detected in the same linkage group which influenced 1000 seed weight, number of pods per plant, percent empty pods, number of seeds plant and biological yield [87]. Clearly, the LG 3 and LG 4 regions appear to have a strong effect on yield under stress. These QTLs appear to be pleiotropic for a number of traits and selection for those with the highest heritability should assist crop improvement for drought stress response [87]. Furthermore, one genomic region concealing QTL for selected drought tolerant traits has been identified and introgressed into three popular chickpea varieties by using a marker-assisted back crossing approach [88]. This attempt will lead to develop improved lines with greater drought tolerance.

3. Heat stress

3.1 Context of heat stress

Adverse effect of heat stress on grain legumes is increasing due to global warming. Chickpea is an important grain legume mostly grown in semi-arid regions which frequently encounters of heat waves that affects crop growth and yield [89]. High temperatures stress in chickpea production is mainly associated with climate change (e.g. summer dominant rainfall zone in Australia) and changes in cropping systems that have shifting chickpea production from cooler region to warmer region (e.g. India) [90, 91]. Both situations, the crop is experiencing high temperature during reproductive stage. Therefore, the effects of heat stress on chickpea growth, development and yield are important to understand by observing agronomic, physiological and biochemical traits to develop high temperature tolerant cultivars.

3.2 Agronomic traits and their relationship with grain yield

Generally, temperature (>30°C) limited yield in cool season legumes such as chickpea, lentil, faba bean and field peas [92–95]. Plant growth, phenology, biomass accumulation and yield are important agronomic traits which depends on the crop ability to withstand or acclimate under abiotic stress [96]. Phenological traits such as days to first flowering, days to 50% flowering and days to crop maturity plays a major role under high temperature. Under stress plants forced to maturity i.e. escaping from heat. Therefore, earliness can be observed through phenological traits. Significant variation in phenology in chickpea under heat stress was observed. Particularly, days to 50% flowering was delayed and days to crop maturity was hastened due to requirement of thermal time (growing degree days °C) to attain any developmental stage [92]. Furthermore, grain yield under heat stress was negatively associated with phenology. To eliminate heat escape, classification of genotypes based on maturity (short, medium and long duration) and stress tolerance index would helpful to identify the genotypes that could be used for future breeding [92, 97].

Plant height, plant width, biomass accumulation, pod number, filled pod number, seed number per plant and grain yield are also plays significant role under high temperature. Plant height and width was affected under heat stress as well as biomass accumulation, pod number and seed weight [92, 98]. Generally, high temperature reduces the duration of plant developmental stages and carbon assimilation process within the plant, resulting low biomass production and reduces source-sink activity (yield) [99]. The most affected yield traits in chickpea are pod number per plant and harvest index. Similar findings have been found in lentil [100].

3.3 Physiological traits and their relationship with grain yield

The difference between heat tolerant and sensitive genotypes of legumes can be identified on the basis of physiological traits such as photosynthetic activity, canopy temperature, CO_2 uptake and membrane thermostability. There are limited research findings available in cool season grain legumes for physiological traits and their relationship with heat stress. Generally, high temperature can negatively affect photosynthesis. The sensitive chickpea genotype at 40/30°C reduced chlorophyll content with a symptom of chlorosis leaves. The symptom of chlorosis in heat stressed plants is common and it was evident in mung bean [101]. Due to inhibition of chlorophyll synthesis, the chlorophyll content may be affected under stress. However, the tolerant chickpea genotype maintained greater chlorophyll content and photochemical efficiency than sensitive genotypes [101] which correlated with yield reduction in sensitive genotypes [102]. This is a clear example to screen genetic material based on photosynthetic activity for heat tolerance.

High temperature affects membrane structure and function. Stress injury can be regulated by loss of membrane integrity and leakage of ions from cells [103]. Therefore, monitoring the function of membrane through electrolyte leakage has been used to screen thermostability under high temperature. The effects of heat stress on the function of membrane has been studied in legumes. Cell membrane thermostability and its correlation with sensitivity was observed by Srinivasan, Takeda [104] and chickpea is the most sensitive crop to high temperature. Similarly, membrane thermostability had linked with sensitivity in chickpea, lentil and faba bean [105]. In lentil, high temperature increased electrolyte leakage due to membrane damage [106]. Thus, membrane damage can be considered as an indicator of heat stress tolerance in legumes and can be used as a physiological trait for screening. Furthermore, Awasthi et al., [107] suggested that drought or heat and combined stresses decreased cellular oxidising ability, stomatal conductance, PSII function and leaf chlorophyll content in chickpea.

Transpiration is the main reason of changes in leaf temperature due to abiotic stresses [105, 108]. Canopy temperature can be sustained through transpiration by open stomata and maintained cool canopy. It was confirmed as an important physiological trait in wheat for high temperature tolerance [109]. In addition to that, canopy temperature depression (CTD) is an indicator of the difference between plant canopy and air temperature. Since the plant closes stomata for certain period due to stress, this will change canopy temperature [33]. Canopy temperature was observed and CTD was calculated using a large set of chickpea genotypes in the field for heat tolerance [97]. Therefore, CTD can be used as a screening method in legumes [97] as it is widely used in cereals [110].

3.4 Biochemical traits and their relationship with grain yield

Soluble sugars play an essential role in plant metabolism, energy production, sugar sensing and signalling in the plant. Generally, biochemical responses in plants observe in the reproductive stage i.e. final stage of grain legumes which involves synthesis of carbohydrates, proteins and lipids in seeds [111]. A positive correlation was found between seed dry weight and sucrose synthase activity under water stress in chickpea [112]. In chickpea, starch metabolism in the leaves affects sucrose availability in the developing seeds and the activities of enzymes related to these metabolic pathways were assessed by Awasthi et al. [107]. Starch concentration, the starch synthesising enzyme were increased under heat-stressed chickpea plants than non-stressed plants. In the seed, the activity of enzyme was inhibited under heat stress. Sucrose in leaves and seeds, sucrose synthase in leaves and seeds and starch phosphorylase in seeds had strong correlation with seed weight per plant and biomass production under heat stress [107]. In lentil, sucrose concentration, sucrose phosphate synthase activity in leaves was significantly low under stressed condition compared non-stressed conditions, finally which influenced seed yield. Similar results were found in wheat under stress and furthermore wheat flour quality was affected [113]. This research confirms that biochemical traits are also plays a role in grain yield under heat stress.

3.5 Pollen as a trait and its relationship with pod set

In legumes, reproductive stage is known to be more sensitive to high temperature than vegetative stage. In reproductive stage, pre-anthesis, anthesis and post-anthesis are important developmental stages which are considered to be sensitive stages among flowering. Heat stress affects reproductive development in chickpea [114], lentil [115], common bean [116] and soybean [117]. Male (anther, pollen) and female organs (stigma-style, ovary) of flowers are severally affected by heat stress (≥30°C) associated with abscission of flower buds, flowers and pods, leading to significant yield loss [118, 119]. Recent findings in legumes revealed that pollen grains are more susceptible to high temperature, thus; chickpea (35/20°C) [114], lentil (35/25°C) [115], soybean (34/24°C) [117]. The effects of high temperature stress in cereals such as wheat, maize and barley also suggested that male reproductive organs might be more sensitive than female organs [120]. Pollen sterility depends on tapetum (anther tissue) and pollen mother cell for pollen formation [121]. Under heat stress, pollen grain fertility is also associated with sucrose content in leaves and anthers [102, 122]. Under stress, pollen tube growth rate also plays an important role [123]. Pollen sterility due to heat stress in cool season legumes such as chickpea and lentil affects pod set and yield [124, 125]. Generally, pollen morphology, in vitro pollen germination and stigma length is a good predictor for hybridization success in lentil [126]. Therefore, analysis of pollen viability and pod set under high temperature is a trait for the study of high temperature tolerance in chickpea.

3.6 Genomic regions linked to heat tolerance

In the recent years, molecular markers have been used to understand quantitatively important traits and markers linked to genomic regions (Quantitative Trait Loci—QTL/genes). Through association analysis, markers have been used to identify the QTLs/genes linked with economically important traits [85]. Association analysis has been clearly comprehended in cereals such as wheat [127, 128], maize [129], barley [130], and sorghum [131]. In winter wheat, QTLs and markers associated with seedling heat tolerance was studied which is useful for early planting and dual-purpose wheat breeding in United States [128]. In legumes, very limited reports are available for association studies particularly under abiotic stress. Association analysis was conducted to identify genomic regions linked to heat tolerance in chickpea. In this study, 107 DArT markers were linked with 11 traits under heat stressed and non-stressed conditions. Four agronomic traits such as total number of pods, filled pods, number of seeds and grain yield were linked to DArT markers under heat stressed and non-stressed conditions. Many significant mapped markers associated in genomic regions in the current study align with previously reported QTLs that influence traits such as plant height, plant width, pod number and grain yield. Therefore, this study identified genomic regions associated with heat tolerance in chickpea and identification of the genes or QTLs linked to this response is the obvious next step [124]. Thudi et al., [88] suggested that 100-seed weight is an important trait linked with 70 significant markers under drought and heat stresses. Other traits such as transpiration efficiency, plant height, root dry weight, pods per plant and yield had significant association under stress. This information can be used for further validation and provide base knowledge to develop tolerant chickpea varieties for drought and heat. Recently, QTLs responsible for traits such as number of filled pods per plot, total number of seeds per plot, grain yield per plot and % pod set were found [132]. The markers linked to QTLs under heat stress will enable marker assisted breeding in the future and assist to understand the mechanism of heat tolerance.

4. Cold stress

4.1 Context of cold stress

Chickpea experiences low temperature (0–12°C) in north India and Australia as a spring crop. It is grown in Western Asia and North Africa and Europe as a winter crop which experiences freezing temperature (down to −10°C) [133, 134]. Low temperature has negative impact on yield and 15–20% of yield loss was estimated [15]. Temperature below 15°C during flowering leads to flower and pod abortion then poor yield [135, 136]. Therefore, the effects of low and freezing temperatures in vegetative and reproductive stages need to study by observing agronomic, physiological and biochemical traits to develop cold temperature tolerant cultivars.

4.2 Agronomic traits and their relationship with grain yield

Chickpea crop is affected by low temperature during flowering (<10°C) in India and Australia. Therefore, low temperature tolerant cultivars or varieties having ability to set pod at <8°C is needed [15]. Low temperature during vegetative stage produced poor vegetative growth, biomass production and yield in north India. While in middle-eastern regions needs chilling tolerance from vegetative to reproductive stage. Germination and early vegetative period are important growth stages under low temperature tolerance [137]. Screening methods for chilling temperatures ranges—1–7°C with pod set developed by ICRISAT [138]. For freezing temperature during early vegetative stage or seedling stage plants were scored using scale 1–9 [139]. Toker [140] found that selected wild *Cicer* species had more freezing tolerance than well-known cold tolerant cultivars.

The effect of chilling range of temperature depends on the phenological stage of the crop. At germination, it causes poor crop establishment and vegetative stage, it results poor crop growth and dry matter accumulation which leads to reduced source-sink production and potential yield [141]. Generally, winter sown chickpea produces more vegetative structure and intercepts with photosynthetically active radiation (PAR) with sufficient dry matter production [142]. Improved exploitation of PAR increases total biomass production and sustains the harvest index similar to spring sown crop [143]. At low temperature, pod set was observed however, seed development was affected. Day and night temperatures play an important role in seed development for cold tolerance [144]. Kanouni et al. [134] suggested that seed yield and had significant correlation with number of secondary branches for freezing temperature tolerance. Correlation between 100 seed weight and cold tolerance rate was also significant. Those traits are important agronomic traits for cold and freezing tolerance.

4.3 Physiological and biochemical traits and their relationship with grain yield

Generally, cold stress causes damage to photosystem II and reduces the stability of chloroplast membranes and photosynthesis. It is confirmed in soybean [145]. Cold stress also results in membrane integrity leading to solute leakage. Under freezing temperature, ice formation in plant tissues is a major injury leads to dehydration [146]. Studies at biochemical level have been reported different changes in crops such as soluble carbohydrate content, soluble protein content and degree of fatty acid content in the shoot cells that perform to complement cold acclimation [147–149]. Elevation of abscisic acid (ABA) and calcium is also associated with cold acclimation [150, 151]. Effect of cold acclimation (1–7°C) on chickpea early vegetative stage was observed by [152]. At 4°C, the electrolyte leakage increased and triphenyl tetrazolium (TTC) content was decreased, finally relative growth rate of root also stopped. ABA content increased in cold acclimated seedling than non-acclimated. The non-acclimated seedlings exposed to 4°C with ABA application showed cold acclimation like response by increasing soluble carbohydrate content. This is an evidence for cold tolerance influenced by ABA content [153]. Later, the influence of ABA application on 40 days old plants and its biomass production and yield was observed in the field [154]. The application of ABA reduces the low-temperature induced biomass production and responsible for yield improvements due to additional photosynthesis [154]. Therefore, ABA application confers cold tolerance in chickpea. Similar effects have been reported in soybean and other grain crops such as rice and maize [155].

4.4 Pollen as a trait and its relationship with pod set

The failure of pod set at low temperature (15–20°C for day and <8°C for night) was observed in chickpea [156] which shows the sensitivity of reproductive organs to cold. At low temperature (15/0°C; 15/5°C) pollen viability and pollen tube growth was affected and pod set was reduced [157]. Clarke and Siddique [158] confirmed the results of Srinivasan, Saxena [157] and justified that low temperature stress during pollen development caused changes in pollen viability and following seed set. It shows the sensitivity of meiosis in the male gametophyte under cold stress. Therefore, pollen viability and its relationship with pod set, seed development and yield are an indirect trait for cold tolerance in chickpea.

4.5 Genomic regions linked to cold tolerance

Limited information is available on the chromosomal locations on the genes and QTLs for cold tolerance in cool season legumes. A study on QTL mapping for winter hardiness in lentil found that cold tolerance is a multi-genic trait. Seven QTLs were identified for winter hardiness in lentil [159]. Recently association analysis of cold and drought stresses along with Ascochyta blight in chickpea was studied using amplified fragment length polymorphism (AFLP) markers [160]. The AFLP markers linked to genes controlling stem number, first effective raining after sowing to 50% flowering, maturity, partial tolerance to Ascochyta blight, 100 seed weight and yield were identified. Identified markers for cold and drought tolerance were not significant. Furthermore, there is no evidence for QTLs reported for cold tolerance in chickpea. Therefore, finding of molecular markers associated with genes controlling different traits under stress could increase the efficiency of marker assisted breeding for abiotic stresses [160].

5. Conclusion and recommendation

Abiotic stresses limit chickpea productivity by affecting its growth and development. Drought, high and low temperatures generally have negative effects on reproductive stage. Freezing temperatures have a negative impact on crop establishment. High and low temperatures affect pollen viability, pollen tube growth and pod set where as in drought more field-based research is needed to confirm effect of stress on male and female organs. Therefore, pollen can be used as a trait under temperature stress. But, correlation between pollen viability and grain yield needs to be comprehended. Physiological and biochemical traits are also affected by abiotic stresses. The effects of combined stresses on physiological and biochemical traits should also be explored. However, more studies are required in chickpea to better understand the relationships among economically important traits and yield and their influence on grain quality under stress conditions. In addition, QTLs for several drought tolerance traits have been identified that can be targeted for molecular breeding. In this context, QTLs for heat and cold tolerance traits have limited information. In future identifying QTLs that explains significant variation in pollen viability under stress and linked markers would accelerate the breeder's interest.

Author details

Peter Kaloki*, Viola Devasirvatham and Daniel K.Y. Tan
Faculty of Science, School of Life and Environmental Sciences, Plant Breeding Institute, Sydney Institute of Agriculture, The University of Sydney, Cobbitty, NSW, Australia

*Address all correspondence to: p.kaloki@ymail.com

References

[1] Parthasarathy Rao P, Birthal PS, Bhagavatula S, Bantilan MCS. Chickpea and Pigeonpea Economies in Asia: Facts, Trends and Outlook. Andhra Pradesh, India: International Crops Research Institute for the Semi-Arid Tropics; 2010

[2] FAOSTAT. In: Kashiwagi J, Krishnamurthy L, Gaur PM, Upadhyaya HD, Varshney RK, Tobita S, editors. Traits of Relevance to Improve Yield Under Terminal Drought Stress in Chickpea (*C. arietinum* L): Field Crops Research; 2012. pp. 88-95

[3] Gunes A, Inal A, Adak MS, Bagci EG, Cicek N, Araslan F. Effect of drought stress implemented at pre- or post-anthesis stage on some physiological parameters as screening criteria in chickpea cultivars. Journal of Plant Physiology. 2008;**55**(1):59-67

[4] Boyer JS. Plant productivity and environment. Science. 1982;**218**:443-448

[5] Varshney R, Thudi M, Nayak S, Gaur P, Kashiwagi J, Krishnamurthy L, et al. Genetic dissection of drought tolerance in chickpea (*Cicer arietinum* L.). Theoretical and Applied Genetics. 2013;**127**:1-18

[6] IPCC. In: Solomon S, Qin D, Manning M, Chen Z, Marquis M, Averyt KB, Tignor M, Miller HL, editors. Climate Change 2007: The Physical Science Basis. Cambridge, United Kingdom and New York, NY, USA: Cambridge University Press; 2007

[7] Jha UC, Chaturvedi SK, Bohra A, Basu PS, Khan MS, Barh D. Abiotic stresses, constraints and improvement strategies in chickpea. Plant Breeding. 2014;**133**(2):163-178

[8] Jackson P, Robertson M, Cooper M, Hammer G. The role of physiological understanding in plant breeding: From a breeding perspective. Field Crops Research. 1996;**49**(1):11-37

[9] Reynolds M, Trethowan R, van Ginkel M, Rajaran S. Application of physiology in wheat breeding. In: Reynolds M, Ortiz-Monasterio JI, McNab A, editors. Application of Physiology in Wheat Breeding. Mexico: CIMMYT; 2001

[10] Quisenberry JE. The interface between plant physiology and genetics. In: Mauney JR, Stewart JM, editors. Cotton Physiology. Vol. 1. Memphis, Tennessee: The Cotton Foundation; 1986. pp. 633-639

[11] Innes NL. The contribution from conventional plant-breeding. Proceedings of the Royal Society of Edinburgh Section B—Biological Sciences. 1992;**99**:1-10

[12] Simmonds NW. Genotype (G), environment (E) and GE components of crop yields. Experimental Agriculture. 1981;**17**(4):355-362

[13] Araus JL, Slafer GA, Royo C, Serret MD. Breeding for yield potential and stress adaptation in cereals. Critical Reviews in Plant Sciences. 2008;**27**(6):377-412

[14] Porch TG. Application of stress indices for heat tolerance screening of common bean. Journal of Agronomy and Crop Science. 2006;**192**(5):390-394

[15] Chaturvedi S, Mishra D, Vyas P, Mishra N. Breeding for cold tolerance in chickpea. Trends in Biosciences. 2009;**2**(2):1-4

[16] Fischer RA. Optimizing the use of water and nitrogen through breeding of crops. Plant and Soil. 1981;**58**(1-3):249-278

[17] Trethowan RM. Defining a genetic ideotype for crop improvement. In: Fleury D, Whitford R, editors. Crop Breeding: Methods and Protocols. New York, NY: Springer; 2014. pp. 1-20

[18] Richards RA. Physiological traits used in the breeding of new cultivars for water-scarce environments. Agricultural Water Management. 2006;**80**(1-3):197-211

[19] Topal A, Aydın C, Akgün N, Babaoglu M. Diallel cross analysis in durum wheat (*Triticum durum* Desf.): Identification of best parents for some kernel physical features. Field Crops Research. 2004;**87**(1):1-12

[20] Richards RA. The effect of dwarfing genes in spring wheat in dry environments. 1. Agronomic characteristics. Australian Journal of Agricultural Research. 1992;**43**(3):517-527

[21] Trethowan R, Reynolds M. Drought resistance: Genetic approaches for improving productivity under stress. In: Wheat Production in Stressed Environments. Dordrecht: Springer; 2007. pp. 289-299

[22] Siddique KHM, Tennant D, Perry MW, Belford RK. Water-use and water-use efficiency of old and modern wheat cultivars in a Mediterranean-type environment. Austalian Journal of Agricultural Research. 1990;**41**(3):431-447

[23] Blum A. Drought adaptation in cereal crops. In: Ribaut J, editor. Drought Adaptation in Cereals. New York: Food Products Press; 2006

[24] Monneveux P, Ribaut J-M. Secondary traits for drought tolerance improvement in cereals. In: Ribaut J-M, editor. Drought Adaptation in Cereals. New York: Food Products Press; 2006

[25] Tuberosa R. Phenotyping drought-stressed crops: Key concepts, issues, and approaches. In: Monneveux P, Ribaut J, editors. Drought Phenotyping in Crops: From Theory to Practice. Mexico: CGIAR Generation Challenge Programme; 2011. pp. 3-24

[26] Montes JM, Melchinger AE, Reif JC. Novel throughput phenotyping platforms in plant genetic studies. Trends in Plant Science. 2007;**12**(10):433-436

[27] Chenu K, Cooper M, Hammer GL, Mathews KL, Dreccer MF, Chapman SC. Environment characterization as an aid to wheat improvement: Interpreting genotype-environment interactions by modelling water-deficit patterns in North-Eastern Australia. Journal of Experimental Botany. 2011;**62**(6):1743-1755

[28] Upadhyaya HD, Kashiwagi J, Varshney RK, Gaur PM, Saxena KB, Krishnamurthy L, et al. Phenotyping chickpeas and pigeonpeas for adaptation to drought. In: Monneveu P, Ribau J-M, editors. Drought Phenotyping in Crops: From Theory to Practice: Generation Challenge Program (GCP); 2011

[29] Peñuelas J, Filella I. Visible and near-infrared reflectance techniques for diagnosing plant physiological status. Trends in Plant Science. 1998;**3**(4):151-156

[30] Jackson RD, Reginato RJ, Idso SB. Wheat canopy temperature: A practical tool for evaluating water requirements. Water Resources Research. 1977;**13**(3):651

[31] Mariano CC, Pietragalla J, Reynolds M. Canopy temperature and plant water relations traits. In: Reynolds M, Pask A, Mullan D, editors. Physiological Breeding I: Interdisciplinary Approaches to Improve Crop Adaptation. Mexico: CIMMYT; 2012. pp. 60-68

[32] Pietragalla J. Canopy temperature. In: Pask AJD, Pietragalla J, Mullan DM,

Reynolds MP, editors. Physiological Breeding II: A Field Guide to Wheat Phenotyping. Mexico: CIMMYT; 2012

[33] Kashiwagi J, Krishnamurthy L, Upadhyaya HD, Gaur PM. Rapid screening technique for canopy temperature status and its relevance to drought tolerance improvement in chickpea. Journal of SAT Agricultural Research. 2008;**6**:1-4

[34] Moran MS, Inoue Y, Barnes EM. Opportunities and limitations for image-based remote sensing in precision crop management. Remote Sensing of Environment. 1997;**61**(3):319-346

[35] Bauer ME. Spectral inputs to crop identification and condition assessment. Proceedings of the IEEE. 1985;**73**(6):1071-1085

[36] Shanahan JF, Schepers JS, Francis DD, Varvel GE, Wilhelm WW, Tringe JM, et al. Use of remote-sensing imagery to estimate corn grain yield. Agronomy Journal. 2001;**93**(3):583-589

[37] Korobov RM, Railyan VY. Canonical correlation relationships among spectral and phytometric variables for 20 winter-wheat fields. Remote Sensing of Environment. 1993;**43**(1):1-10

[38] Tucker CJ. Red and photographic infrared linear combinations for monitoring vegetation. Remote Sensing of Environment. 1979;**8**(2):127-150

[39] Douglas Ramsey R, Falconer A, Jensen JR. The relationship between NOAA-AVHRR NDVI and ecoregions in Utah. Remote Sensing of Environment. 1995;**53**(3):188-198

[40] Teillet PM. An algorithm for the radiometric and atmospheric correction of AVHRR data in the solar reflective channels. Remote Sensing of Environment. 1992;**41**(2):185-195

[41] Bonhomme R. Beware of comparing RUE values calculated from PAR vs solar radiation or absorbed vs intercepted radiation. Field Crops Research. 2000;**68**(3):247-252

[42] Zhang L, Zhang S, van der Werf W, Bastiaans L, Li B, Spiertz JHJ. Light interception and utilization in relay intercrops of wheat and cotton. Field Crops Research. 2008;**107**(1):29-42

[43] Johnson MVV, Kiniry JR, Burson BL. Ceptometer deployment method affects measurement of fraction of intercepted photosynthetically active radiation. Agronomy Journal. 2010;**102**(4):1132-1137

[44] Thomson BD, Siddique KHM. Grain legume species in low rainfall Mediterranean-type environments II. Canopy development, radiation interception, and dry-matter production. Field Crops Research. 1997;**54**(2):189-199

[45] Kiniry JR, Simpson CE, Schubert AM, Reed JD. Peanut leaf area index, light interception, radiation use efficiency, and harvest index at three sites in Texas. Field Crops Research. 2005;**91**(2):297-306

[46] Andrade FH, Calviño P, Cirilo A, Barbieri P. Yield responses to narrow rows depend on increased radiation interception. Agronomy Journal. 2002;**94**(5):975-980

[47] Cartelat A, Agati G, Moya I, Cerovic ZG, Goulas Y, Meyer S, et al. Optically assessed contents of leaf polyphenolics and chlorophyll as indicators of nitrogen deficiency in wheat (*Triticum aestivum* L.). Field Crops Research. 2005;**91**(1):35-49

[48] Li R-h, P-g G, Michael B, Stefania G, Salvatore C. Evaluation of chlorophyll content and fluorescence parameters as indicators of drought tolerance in barley. Agricultural Sciences in China. 2006;**5**(10): 751-757

[49] Liu YB, Zhang TG, Wang J. Photosynthesis and metabolite levels in dehydrating leaves of Reaumuria soongorica. Acta Biologica Cracoviensia Series Botanica. 2008;**50**(1):19-26

[50] Cenkci S, Ciğerci İH, Yıldız M, Özay C, Bozdağ A, Terzi H. Lead contamination reduces chlorophyll biosynthesis and genomic template stability in *Brassica rapa* L. Environmental and Experimental Botany. 2010;**67**(3):467-473

[51] Jangpromma N, Songsri P, Thammasirirak S, Jaisil P. Rapid assessment of chlorophyll content in sugarcane using a SPAD chlorophyll meter across different water stress conditions. Asian Journal of Plant Sciences. 2010;**9**(6):368-374

[52] Ahmed SU. Effects of soil water deficit on leaf nitrogen, chlorophylls and spad chlorophyll meter reading on growth stages of soybean. Bangladesh Journal of Botany. 2011;**40**(2):171-175

[53] Markwell J, Osterman JC, Mitchell JL. Calibration of the Minolta SPAD-502 leaf chlorophyll meter. Photosynthesis Research. 1995;**46**(3):467-472

[54] Lynch J. Root architecture and plant productivity. Plant Physiology. 1995;**109**(1):7-13

[55] Ludlow MM, Muchow RC. A critical evaluation of traits for improving crop yields in water-limited environments. Advances in Agronomy. 1990;**43**(C):107-153

[56] Kashiwagi J, Krishnamurthy L, Crouch JH, Serraj R. Variability of root length density and its contributions to seed yield in chickpea (*Cicer arietinum* L.) under terminal drought stress. Field Crops Research. 2006;**95**(2):171-181

[57] Varshney RK, Tripathi S, Rathore A, Jukanti AK, Jayalakshmi V, Vemula A, et al. Genetic dissection of drought tolerance in chickpea (*Cicer arietinum* L.). Theoretical and Applied Genetics. 2014;**127**(2):445-462

[58] Bramley H, Turner NC, Turner DW, Tyerman SD. Roles of morphology, anatomy, and aquaporins in determining contrasting hydraulic behavior of roots. Plant Physiology. 2009;**150**(1):348-364

[59] Zaman-Allah M, Jenkinson DM, Vadez V. A conservative pattern of water use, rather than deep or profuse rooting, is critical for the terminal drought tolerance of chickpea. Journal of Experimental Botany. 2011;**62**(12):4239-4252

[60] Passioura JB. Grain yield, harvest index, and water use of wheat. Journal of the Australian Institute of Agricultural Science. 1977;**43**:117-120

[61] Krishnamurthy L, Vadez V, Varshney RK, Kashiwagi J, Tobita S, Ito O, et al. Variation in carbon isotope discrimination and its relationship with harvest index in the reference collection of chickpea germplasm. Functional Plant Biology. 2013;**40**(12):1350-1361

[62] Farquhar GD, Ehleringer JR, Hubick KT. Carbon isotope discrimination and photosynthesis. Annual Review of Plant Biology. 1989;**40**(1):503-537

[63] Farquhar GD, Richards RA. Isotopic composition of plant carbon correlates with water-use efficiency of wheat genotypes. Australian Journal of Plant Physiology. 1984;**11**(6):539

[64] Ali MH, Talukder MSU. Increasing water productivity in crop production—A synthesis. Agricultural Water Management. 2008;**95**(11):1201-1213

[65] Condon AG, Richards RA, Rebetzke GJ, Farquhar GD. Improving intrinsic water-use efficiency and crop yield. Crop Science. 2002;**42**(1):122-131

[66] Rebetzke GJ, Condon AG, Richards RA, Farquhar GD. Selection for reduced carbon isotope discrimination increases aerial biomass and grain yield of rainfed bread wheat. Crop Science. 2002;**42**(3):739-745

[67] Condon AG, Richards RA, Rebetzke GJ, Farquhar GD. Breeding for high water-use efficiency. Journal of Experimental Botany. 2004;**55**(407):2447-2460

[68] Reynolds M, Tuberosa R. Translational research impacting on crop productivity in drought-prone environments. Current Opinion in Plant Biology. 2008;**11**(2):171-179

[69] Bahadur A, Chatterjee A, Kumar R, Singh M, Naik PS. Physiological and biochemical basis of drought tolerance in vegetables. Vegetable Science. 2011;**38**(1):1-16

[70] Shohael AM, Ali MB, Yu KW, Hahn EJ, Islam R, Paek KY. Effect of light on oxidative stress, secondary metabolites and induction of antioxidant enzymes in *Eleutherococcus senticosus* somatic embryos in bioreactor. Process Biochemistry. 2006;**41**(5):1179-1185

[71] Zhu Z, Liang Z, Han R. Saikosaponin accumulation and antioxidative protection in drought-stressed *Bupleurum chinense* DC. Plants. Environmental and Experimental Botany. 2009;**66**(2):326-333

[72] Iturbe-Ormaetxe I, Becana M. Oxidative damage in pea plants exposed to water deficit or paraquat. Plant Physiology. 1998;**116**(1):173-181

[73] Mittler R. Oxidative stress, antioxidants and stress tolerance. Trends in Plant Science. 2002;7(9):405-410

[74] Smirnoff N. The role of active oxygen in the response of plants to water-deficit and desiccation. New Phytologist. 1993;**125**(1):27-58

[75] Bright J, Desikan R, Hancock JT, Weir IS, Neill SJ. ABA-induced NO generation and stomatal closure in arabidopsis are dependent on H_2O_2 synthesis. The Plant Journal. 2006;**45**(1):113-122

[76] Noctor G, Foyer CH. A re-evaluation of the ATP : NADPH budget during C-3 photosynthesis: A contribution from nitrate assimilation and its associated respiratory activity. Journal of Experimental Botany. 1998;**49**(329):1895-1908

[77] Quan LJ, Zhang B, Shi WW, Li HY. Hydrogen peroxide in plants: A versatile molecule of the reactive oxygen species network. Journal of Integrative Plant Biology. 2008;**50**(1):2-18

[78] Foreman J, Jones JDG, Davies JM, Dolan L, Demidchik V, Bothwell JHF, et al. Reactive oxygen species produced by NADPH oxidase regulate plant cell growth. Nature. 2003;**422**(6930):442-446

[79] Nordborg M, Tavaré S. Linkage disequilibrium: What history has to tell us. Trends in Genetics: TIG. 2002;**18**(2):83-90

[80] Zhu C, Gore M, Buckler E, Yu J. Status and prospects of association mapping in plants. Plant Genome. 2008;**1**:5-20

[81] Yu J, Buckler ES. Genetic association mapping and genome organization of maize. Current Opinion in Biotechnology. 2006;**17**(2):155-160

[82] Malosetti M, van der Linden CG, Vosman B, van Eeuwijk FA. A mixed-model approach to association mapping using pedigree information with an illustration of resistance to *Phytophthora infestans* in potato. Genetics. 2007;**175**(2):879-889

[83] Thornsberry JM, Goodman MM, Doebley J, Kresovich S, Nielsen D, Buckler ES. Dwarf8 polymorphisms

associate with variation in flowering time. Nature Genetics. 2001;**28**(3):286-289

[84] Wang J, McClean PE, Lee R, Goos RJ, Helms T. Association mapping of iron deficiency chlorosis loci in soybean (*Glycine max* L. Merr.) advanced breeding lines. Theoretical and Applied Genetics. 2008;**116**(6):777-787

[85] Gowda SJM, Radhika P, Mhase LB, Jamadagni BM, Gupta VS, Kadoo NY. Mapping of QTLs governing agronomic and yield traits in chickpea. Journal of Applied Genetics. 2011;**52**:9-21

[86] Rehman AU, Malhotra RS, Bett K, Tar'an B, Bueckert R, Warkentin TD. Mapping QTL associated with traits affecting grain yield in Chickpea (*Cicer arietinum* L.) under terminal drought stress. Crop Science. 2011;**51**(2):450-463

[87] Hamwieh A, Imtiaz M, Malhotra RS. Multi-environment QTL analyses for drought-related traits in a recombinant inbred population of chickpea (*Cicer arietinum* L.). Theoretical and Applied Genetics. 2013;**126**(4):1025-1038

[88] Thudi M, Upadhyaya HD, Rathore A, Gaur PM, Krishnamurthy L, Roorkiwal M, et al. Genetic dissection of drought and heat tolerance in chickpea through genome-wide and candidate gene-based association mapping approaches. PLoS One. 2014;**9**(5):e96758

[89] Jumrani K, Bhatia VS. Impact of elevated temperatures on growth and yield of chickpea (*Cicer arietinum* L.). Field Crops Research. 2014;**164**:90-97

[90] Gowda CLL, Rao PP, Tripathi S, Gaur PM, Deshmukh RB. Regional shift in chickpea production in India. In: Milestones in Food Legumes Research. Kanpur, India: Indian Institute of Pulses Research; 2009. pp. 21-35

[91] Anwar MR, O'Leary G, Brand J, Redden RJ. Crop simulation model for yield prediction. In: Yadav SS, Redden RJ, Chen W, Sharma B, editors. Chickpea Breeding and Management. UK: CABI; 2007. pp. 575-597

[92] Krishnamurthy L, Gaur P, Basu P, Chaturvedi S, Tripathi S, Vadez V, et al. Large genetic variation for heat tolerance in the reference collection of chickpea (*Cicer arietinum* L.) germplasm. Plant Genetic Resources. 2011;**9**(1):59-69

[93] Bhandari K, Sharma KD, Rao BH, Siddique KH, Gaur P, Agrawal SK, et al. Temperature sensitivity of food legumes: A physiological insight. Acta Physiologiae Plantarum. 2017;**39**(3):68

[94] Bishop J, Potts SG, Jones HE. Susceptibility of faba bean (*Vicia faba* L.) to heat stress during floral development and anthesis. Journal of Agronomy and Crop Science. 2016;**202**(6):508-517

[95] Jiang Y, Lahlali R, Karunakaran C, Kumar S, Davis AR, Bueckert RA. Seed set, pollen morphology and pollen surface composition response to heat stress in field pea. Plant, Cell & Environment. 2015;**38**(11):2387-2397

[96] Prasad P, Staggenborg S, Ristic Z. Impacts of drought and/or heat stress on physiological, developmental, growth, and yield processes of crop plants. In: Response of Crops to Limited Water: Understanding and Modeling Water Stress Effects on Plant Growth Processes. American Society of Agronomy, Crop Science Society of America and Soil Science Society of America; 2008 (response of crops). pp. 301-355

[97] Devasirvatham V, Gaur P, Raju T, Trethowan R, Tan D. Field response of chickpea (*Cicer arietinum* L.) to high temperature. Field Crops Research. 2015;**172**:59-71

[98] Upadhyaya HD, Dronavalli N, Gowda C, Singh S. Identification and evaluation of chickpea germplasm for

tolerance to heat stress. Crop Science. 2011;**51**(5):2079-2094

[99] Harding SA, Guikema JA, Paulsen GM. Photosynthetic decline from high temperature stress during maturation of wheat: II. Interaction with source and sink processes. Plant Physiology. 1990;**92**(3):654-658

[100] Kumar J, Kant R, Kumar S, Basu P, Sarker A, Singh N. Heat tolerance in lentil under field conditions. Legume Genomics and Genetics. 2016;7:1-11

[101] Kumar S, Kaur R, Kaur N, Bhandhari K, Kaushal N, Gupta K, et al. Heat-stress induced inhibition in growth and chlorosis in mungbean (*Phaseolus aureus* Roxb.) is partly mitigated by ascorbic acid application and is related to reduction in oxidative stress. Acta Physiologiae Plantarum. 2011;**33**(6):2091

[102] Kaushal N, Awasthi R, Gupta K, Gaur P, Siddique KHM, Nayyar H. Heat-stress-induced reproductive failures in chickpea (*Cicer arietinum*) are associated with impaired sucrose metabolism in leaves and anthers. Functional Plant Biology. 2013;**40**(12):1334

[103] Salvucci ME, Crafts-Brandner SJ. Mechanism for deactivation of Rubisco under moderate heat stress. Physiologia Plantarum. 2004;**122**(4):513-519

[104] Srinivasan A, Takeda H, Senboku T. Heat tolerance in food legumes as evaluated by cell membrane thermostability and chlorophyll fluorescence techniques. Euphytica. 1996;**88**(1):35-45

[105] Ibrahim HM. Heat stress in food legumes: Evaluation of membrane thermostability methodology and use of infra-red thermometry. Euphytica. 2011;**180**(1):99-105

[106] Barghi SS, Mostafaii H, Peighami F, Zakaria RA, Nejhad RF. Response of in vitro pollen germination and cell membrane thermostability of lentil genotypes to high temperature. International Journal of Agriculture. 2013;**3**(1):13

[107] Awasthi R, Kaushal N, Vadez V, Turner NC, Berger J, Siddique KH, et al. Individual and combined effects of transient drought and heat stress on carbon assimilation and seed filling in chickpea. Functional Plant Biology. 2014;**41**(11):1148-1167

[108] Blum A. Plant Breeding for Stress Environments. Boca Raton, FL: CRC Press; 1988

[109] Munjal R, Rana R. Evaluation of physiological traits in wheat (*Triticum aestivum* L.) for terminal high temperature tolerance. In: Proceedings of the Tenth International Wheat Genetics Symposium, Poestum, Italy. 2003

[110] Rosyara UR, Subedi S, Duveiller E, Sharma RC. The effect of spot blotch and heat stress on variation of canopy temperature depression, chlorophyll fluorescence and chlorophyll content of hexaploid wheat genotypes. Euphytica. 2010;**174**(3):377-390

[111] Weschke W, Panitz R, Sauer N, Wang Q, Neubohn B, Weber H, et al. Sucrose transport into barley seeds: Molecular characterization of two transporters and implications for seed development and starch accumulation. The Plant Journal. 2000;**21**(5):455-467

[112] Turner NC, Furbank RT, Berger JD, Gremigni P, Abbo S, Leport L. Seed size is associated with sucrose synthase activity in developing cotyledons of chickpea. Crop Science. 2009;**49**(2):621-627

[113] Dias AS, Bagulho AS, Lidon FC. Ultrastructure and biochemical traits of bread and durum wheat grains under heat stress. Brazilian Journal of Plant Physiology. 2008;**20**(4):323-333

[114] Devasirvatham V, Gaur PM, Mallikarjuna N, Tokachichu RN, Trethowan RM, Tan DK. Effect of high temperature on the reproductive development of chickpea genotypes under controlled environments. Functional Plant Biology. 2012;**39**(12):1009-1018

[115] Sita K, Sehgal A, Kumar J, Kumar S, Singh S, Siddique KH, et al. Identification of high-temperature tolerant lentil (*Lens culinaris* Medik.) genotypes through leaf and pollen traits. Frontiers in Plant Science. 2017;**8**:744

[116] Gross Y, Kigel J. Differential sensitivity to high temperature of stages in the reproductive development of common bean (*Phaseolus vulgaris* L.). Field Crops Research. 1994;**36**(3):201-212

[117] Jumrani K, Bhatia VS, Pandey GP. Screening soybean genotypes for high temperature tolerance by in vitro pollen germination, pollen tube length, reproductive efficiency and seed yield. Indian Journal of Plant Physiology. 2018;**23**(1):77-90

[118] Nakano H, Kobayashi M, Terauchi T. Sensitive stages to heat stress in pod setting of common bean (*Phaseolus vulgaris* L.). Japanese Journal of Tropical Agriculture. 1998;**42**(2):78-84

[119] Duthion C, Pigeaire A. Seed lengths corresponding to the final stage in seed abortion of three grain legumes. Crop Science. 1991;**31**(6):1579-1583

[120] Sakata T, Higashitani A. Male sterility accompanied with abnormal anther development in plants—Genes and environmental stresses with special reference to high temperature injury. International Journal of Plant Developmental Biology. 2008;**2**(4):42-51

[121] Zinn KE, Tunc-Ozdemir M, Harper JF. Temperature stress and plant sexual reproduction: Uncovering the weakest links. Journal of Experimental Botany. 2010;**61**(7):1959-1968

[122] Kumar S, Thakur P, Kaushal N, Malik JA, Gaur P, Nayyar H. Effect of varying high temperatures during reproductive growth on reproductive function, oxidative stress and seed yield in chickpea genotypes differing in heat sensitivity. Archives of Agronomy and Soil Science. 2013;**59**(6):823-843

[123] Hedhly A, Hormaza JI, Herrero M. Global warming and sexual plant reproduction. Trends in Plant Science. 2009;**14**(1):30-36

[124] Devasirvatham V, Gaur PM, Mallikarjuna N, Raju TN, Trethowan RM, Tan DK. Reproductive biology of chickpea response to heat stress in the field is associated with the performance in controlled environments. Field Crops Research. 2013;**142**:9-19

[125] Sita K, Sehgal A, HanumanthaRao B, Nair RM, Vara Prasad P, Kumar S, et al. Food legumes and rising temperatures: Effects, adaptive functional mechanisms specific to reproductive growth stage and strategies to improve heat tolerance. Frontiers in Plant Science. 2017;**8**:1658

[126] Fratini R, Ruiz ML. Interspecific hybridization in the genus Lens applying in vitro embryo rescue. Euphytica. 2006;**150**(1-2):271-280

[127] Neumann K, Kobiljski B, Denčić S, Varshney R, Börner A. Genome-wide association mapping: A case study in bread wheat (*Triticum aestivum* L.). Molecular Breeding. 2011;**27**(1): 37-58

[128] Maulana F, Ayalew H, Anderson JD, Kumssa TT, Huang W, Ma X-F. Genome-wide association mapping of seedling heat tolerance in winter wheat. Frontiers in Plant Science. 2018;**9**:1-16

[129] Lu Y, Yan J, Guimaraes CT, Taba S, Hao Z, Gao S, et al. Molecular

characterization of global maize breeding germplasm based on genome-wide single nucleotide polymorphisms. Theoretical and Applied Genetics. 2009;**120**(1):93-115

[130] Cockram J, White J, Leigh FJ, Lea VJ, Chiapparino E, Laurie DA, et al. Association mapping of partitioning loci in barley. BMC Genetics. 2008;**9**(1):16

[131] Shehzad T, Iwata H, Okuno K. Genome-wide association mapping of quantitative traits in sorghum (*Sorghum bicolor* (L.) Moench) by using multiple models. Breeding Science. 2009;**59**(3):217-227

[132] Paul PJ, Samineni S, Thudi M, Sajja SB, Rathore A, Das RR, et al. Molecular mapping of QTLs for heat tolerance in chickpea. International Journal of Molecular Sciences. 2018;**19**(8):2166

[133] Toker C, Lluch C, Tejera NA, Serraj R, Siddique KHM. Abiotic stresses. In: Yadav SS, Redden RJ, Chen W, Sharma B, editors. Chickpea Breeding and Management. UK: CABI; 2007

[134] Kanouni H, Farayedi Y, Saeid A, Sabaghpour SH. Stability analyses for seed yield of chickpea (*Cicer arietinum* L.) genotypes in the western cold zone of Iran. Journal of Agricultural Science. 2015;7(5):219

[135] Sharma P, Shekhon HS, JS S, editors. Cold tolerance studies in chickpea (*Cicer arietinum*). In: 4th International Food Legume Research Conference (IFLRC-IV). New Delhi, India: Consultative Group on International Agricultural Research (CGIAR); 2005

[136] Ali M, Kumar S. Chickpea (*Cicer arietinum* L.) research in India: Accomplishments and future strategies. Indian Journal of Agricultural Sciences. 2005;**75**:125-133

[137] Singh KB, Malhotra RS, Halila MH, Knights EJ, Verma MM. Current status and future strategy in breeding chickpea for resistance to biotic and abiotic stresses. Euphytica. 1993;**73**(1): 137-149

[138] ICRISAT. Chickpea kabuli variety ICCV6. In: ICRISAT, editor. Plant Material Description ICRISAT No. 124. 1989

[139] Singh KB, Malhotra RS, Saxena MC. Chickpea evaluation for cold tolerance under field conditions. Crop Science. 1989;**29**(2):282-285

[140] Toker C. Preliminary screening and selection for cold tolerance in annual wild Cicer species. Genetic Resources and Crop Evolution. 2005;**52**(1):1-5

[141] Croser JS, Clarke HJ, Siddique KHM, Khan TN. Low-temperature stress: Implications for chickpea (*Cicer arietinum* L.) improvement. Critical Reviews in Plant Sciences. 2003;**22**(2):185-219

[142] Singh KB, Omar M, Saxena MC, Johansen C. Screening for drought resistance in spring chickpea in the Mediterranean Region*. Journal of Agronomy and Crop Science. 1997;**178**(4):227-235

[143] Siddique K, Sedgley R. Chickpea (*Cicer arietinum* L.), a potential grain legume for South-Western Australia: Seasonal growth and yield. Australian Journal of Agricultural Research. 1986;**37**(3):245-261

[144] Srinivasan A, Johansen C, Saxena NP. Cold tolerance during early reproductive growth of chickpea (*Cicer arietinum* L.): Characterization of stress and genetic variation in pod set. Field Crops Research. 1998;**57**(2):181-193

[145] Heerden PDRV, Krüger GHJ. Photosynthetic limitation in soybean during cold stress.

South African Journal of Science. 2000;**96**(4):201-206

[146] Yadav SK. Cold stress tolerance mechanisms in plants. A review. Agronomy for Sustainable Development. 2010;**30**(3):515-527

[147] Thakur P, Kumar S, Malik JA, Berger JD, Nayyar H. Cold stress effects on reproductive development in grain crops: An overview. Environmental and Experimental Botany. 2010;**67**(3):429-443

[148] Lee T-M, Lur H-S, Chu C. Role of abscisic acid in chilling tolerance of rice (*Oryza sativa* L.) seedlings: II. Modulation of free polyamine levels. Plant Science. 1997;**126**(1):1-10

[149] Bakht J, Bano A, Dominy P. The role of abscisic acid and low temperature in chickpea (*Cicer arietinum*) cold tolerance. II. Effects on plasma membrane structure and function. Journal of Experimental Botany. 2006;**57**(14):3707-3715

[150] Lang V, Mantyla E, Welin B, Sundberg B, Palva ET. Alterations in water status, endogenous abscisic acid content, and expression of rab18 gene during the development of freezing tolerance in *Arabidopsis thaliana*. Plant Physiology. 1994;**104**(4):1341-1349

[151] Knight H, Trewavas AJ, Knight MR. Cold calcium signaling in Arabidopsis involves two cellular pools and a change in calcium signature after acclimation. The Plant Cell. 1996;**8**(3):489-503

[152] Nayyar H, Bains TS, Kumar S. Chilling stressed chickpea seedlings: Effect of cold acclimation, calcium and abscisic acid on cryoprotective solutes and oxidative damage. Environmental and Experimental Botany. 2005;**54**(3):275-285

[153] Nayyar H, Bains T, Kumar S. Low temperature induced floral abortion in chickpea: Relationship to abscisic acid and cryoprotectants in reproductive organs. Environmental and Experimental Botany. 2005;**53**(1):39-47

[154] Bakht J, Bano A, Shafi M, Dominy P. Effect of abscisic acid applications on cold tolerance in chickpea (*Cicer arietinum* L.). European Journal of Agronomy. 2013;**44**:10-21

[155] Frascaroli E, Tuberosa R. Effect of abscisic acid on pollen germination and tube growth of maize genotypes 1. Plant Breeding. 1993;**110**(3):250-254

[156] Saxena NP, Johansen C, editors. Realised yield potential in chickpea and physiological considerations for further genetic improvement. In: Int Congress Plant Physiology. New Delhi: Society for Plant Physiology and Biochemistry; 1990

[157] Srinivasan A, Saxena NP, Johansen C. Cold tolerance during early reproductive growth of chickpea (*Cicer arietinum* L.): Genetic variation in gamete development and function. Field Crops Research. 1999;**60**(3):209-222

[158] Clarke HJ, Siddique KHM. Response of chickpea genotypes to low temperature stress during reproductive development. Field Crops Research. 2004;**90**(2):323-334

[159] Kahraman A, Kusmenoglu I, Aydin N, Aydogan A, Erskine W, Muehlbauer FJ. QTL mapping of winter hardiness genes in lentil. Crop Science. 2004;**44**(1):13

[160] Saeed A, Darvishzadeh R. Association analysis of biotic and abiotic stresses resistance in chickpea (Cicer spp.) using AFLP markers. Biotechnology & Biotechnological Equipment. 2017;**31**(4):698-708